When Critical Multiculturalism Meets Mathematics

When Critical Multiculturalism Meets Mathematics

A Mixed Methods Study of Professional Development and Teacher Identity

Patricia L. Marshall
Jessica T. DeCuir-Gunby & Allison W. McCulloch

ROWMAN & LITTLEFIELD
Lanham • Boulder • New York • London

Published by Rowman & Littlefield
A wholly owned subsidiary of The Rowman & Littlefield Publishing Group, Inc.
4501 Forbes Boulevard, Suite 200, Lanham, Maryland 20706
www.rowman.com

Unit A, Whitacre Mews, 26-34 Stannary Street, London SE11 4AB

British Library Cataloguing in Publication Data

Library of Congress Cataloging-in-Publication Data Available
ISBN 978-1-4758-0848-3 (cloth : alk. paper)
ISBN 978-1-4758-0849-0 (pbk. : alk. paper)
ISBN 978-1-4758-0850-6 (electronic)

∞™ The paper used in this publication meets the minimum requirements of American National Standard for Information Sciences—Permanence of Paper for Printed Library Materials, ANSI/NISO Z39.48-1992.

Printed in the United States of America

This book is dedicated to the talented and courageously curious classroom teachers who participated in our study. Sincere thanks are extended to each of them for accepting our invitation to grapple seriously with big ideas and complex questions about critical multiculturalism and mathematics teaching in contemporary culturally diverse schools.

Contents

List of Figures and Tables

FIGURES

BOX

TABLES

Foreword

For many years I have conducted professional development workshops and seminars for teachers on culturally relevant pedagogy across the nation and the world. Invariably someone raises a question about the applicability of this work for his or her content area or discipline. The question usually goes something like, "Well, I see how this stuff might work in English-language arts, social studies, art or music but I teach math (or science)! Culturally relevant pedagogy and multicultural education doesn't apply to me!" Early in my career I would spend time giving the questioner example after example of ways he or she could use the principles of my work in her teaching, but over the last few years I have taken to responding to the assertions of the neutrality of the field of mathematics (or science) by asking the question, "Tell me how your African American (and Latino) students are doing in mathematics?" The response to that question is predictable: "They are struggling (or failing)." At that point I reply, "So, are you interested in doing something to improve that?"

The point of my questions is that teachers must decide if their allegiance to the "integrity of the discipline" overrides their commitment to the success of all students. If our teachers believe they are merely teaching mathematics rather than teaching students who are learning mathematics, we are likely to encounter resistance at every turn. We will hear that students are either unprepared or under-prepared. We will hear that some students are unmotivated. We will hear that some students don't want to learn. Sometimes we will hear that some students are incapable of learning.

Civil rights leader, mathematician, and author of *Radical Equations: Civil Rights from Mississippi to the Algebra Project* Robert Moses asserted that while literacy (learning to read, write, listen, and speak) was the most important skill for the 1960s era civil rights, today's most pressing civil right is mathematical literacy. Thus, we must ensure that all students, regardless of race, ethnicity, language, or any other perceived differences, have access to high-quality mathematics teaching.

When Critical Multiculturalism Meets Mathematics: A Mixed Methods Study of Professional Development and Teacher Identity is a look at what happens when teachers are willing to engage in work that helps to improve the academic, cultural, and socio-civic success of all students—particularly African American and Latino students. This work is reminiscent of the work that I engaged in over 20 years ago, when I searched for successful teachers of African American students. That work resulted in the publication of *The Dreamkeepers: Successful Teachers of African American Children*. One of the teachers called Ms. Rossi, who I have highlighted in the book, was determined to ensure that her low-income African American 6th graders had access to high-quality mathematics. So, Ms. Rossi scraped the prescribed curriculum and began teaching her students algebra. She scrounged for books, pushed the students to learn the concepts, and reminded them regularly that they were more than capable of mastering the mathematics.

Teachers like Ms. Rossi are made, not born. They work at their craft and understand that good teachers, like good students, are perpetual learners. They take seriously the need to meet students where they are and take them to where they need to be. This volume details the process of "making" good teachers. Marshall, DeCuir-Gunby, and McCulloch share the work of professional development that builds teachers' confidence and willingness to stretch in new ways and in new directions. The work of teaching mathematics for equity and justice is a form of culturally relevant pedagogy.

Culturally relevant pedagogy is a theoretical approach to teaching that is built on three propositions—student learning, cultural competence, and sociopolitical consciousness. Student learning is obviously what students accomplish academically while under the supervision of teachers. The teachers who are culturally relevant are capable of explaining what students know and are able to do when they arrived in their classrooms and what they are capable of doing when the year ends or any point in between. They are the teachers who push students intellectually and raise standards because they believe their students will rise to the heights they set.

Cultural competence involves helping students appreciate their culture of origin while also becoming fluent in at least one other culture. For students of color this typically means being confident in their histories, languages, customs, and cultures while acquiring the skills that allow for success in the mainstream. Cultural competence also applies to White middle-class students. While schools already cater to their culture and experiences, they too need to learn how to navigate an increasingly complex and multicultural world, and that navigation requires cultural competence. But, how can our students become culturally competent when far too many of their teachers are culturally incompetent?

Culturally relevant pedagogy requires that teachers help students engage with the wider community—their neighborhoods, cities, state, nation, and even the world. This sociopolitical consciousness is what I think of as the "so-what factor." We can teach students how to find square roots or solve quadratic equations, but we should also be helping them understand the value of this knowledge and skill beyond the classroom. Sociopolitical consciousness is a way to help students see the value of what they are learning to help them solve problems of living in a diverse society plagued by divisions along racial, ethnic, linguistic, economic, religious, cultural, and other differences.

In addition to sharing the experience of preparing teachers to teach mathematics to all students, this volume takes on the knotty problems associated with the culture of mathematics education. Far too many who work in mathematics education continue to have a perspective that suggests that high-level mathematics learning only occurs among a "certain" group of students. Psychologist Carol Dweck of Stanford University calls this kind of mind-set the "innate ability" perspective and contrasts this perspective with the "effort" one. The innate ability perspective asserts that ability is heritable. You are born with certain ability and capacity and you either have it or you don't. People who take this stance will make comments like, "Well, I was never any good at math so I know my kid won't be." The effort perspective argues that excellent performance is a result of increased effort. Some societies and cultures emphasize this perspective. For instance, in Japan, parents will argue that the reason for the difference between a student who earns a higher score on an exam and the one who struggles is that the higher performer slept an hour less. The implication is that if you want to do well, you study more! This worldview stands in stark contrast to the predominant view in the United States, which suggests that some people are smart and some are not.

Today, STEM (science, technology, engineering, and mathematics) related skills and knowledge are crucial to social, civic, and economic advancement. With an increasingly diverse society, we cannot afford for any of our students to be unprepared in these areas. We need every student to be capable of solving complex problems and accessing the tools of science and mathematics to help create a more equitable and just society. In order to make that happen we need teachers who can foster student excellence in mathematics. Marshall, DeCuir-Gunby, and McCulloch show us that our teachers can learn to meet the needs of all students as they learn mathematics. If we are fortunate, this volume will force us to re-examine how we prepare and support teachers to seek equitable pedagogies in every subject area.

Gloria Ladson-Billings,
Kellner Family Distinguished Chair in Urban Education
University of Wisconsin-Madison
2015

Acknowledgments

The successful completion of the research reported in this book represents the culmination of work by many individuals. Some were principal contributors to the initial idea for the study, whereas others assumed central roles at different points during its middle or late stages. We genuinely appreciate the contributions of all, but for various reasons, the identities and specific roles of some shall remain anonymous.

Initially, appreciation is extended to our editors at Rowman & Littlefield. Our first editor, Nancy Evans, expressed interest in and enthusiasm for the prospectus for this book; however, it was our second editor, Sarah Jubar, who took the ride with us through its complete development.

Our project was funded by a grant from the National Science Foundation. During its span we had the fortune to work with stellar Program Officers, including Monica Mitchell, Paola Sztajn, Angelicque Tucker-Blackmon, John "Spud" Bradley, and David B. Campbell. We would be wholly remiss if we did not acknowledge a collection of advisors, colleagues, and consultants. These individuals either participated in some part of the original formulation and conduct of the study, offered expertise in its critical early stages, contributed to its partial reconceptualization during particularly stressful periods, or simply came through with invaluable advice in the proverbial nick of time. They include George Bright, Patricia Campbell, Beatriz D'Ambrosio, Carl Grant, Jeane Joyner, Gloria Ladson-Billings, Hollylynne Stohl Lee, Martin Simon, Christine Sleeter, Lee Stiff, William Tate, Ron Tzur, and Dorothy White.

This research could never have progressed beyond ideas on paper had it not been for the commitment of the central office administrators and building principals in the school district where the study took place. Similarly, the efforts of our Research Assistants and Data Collectors were immensely vital during every phase of the study.

Finally, while we are indebted to the efforts of many, ultimately we accept that all limitations and missteps in the telling of the story as presented in this book are ours alone.

Patricia L. Marshall
Jessica T. DeCuir-Gunby
Allison W. McCulloch

Chapter 1

Multicultural Education and High-Quality Mathematics Teaching

A question about multicultural education was posed during the oral segment of a comprehensive examination for Shania,[1] an African American woman with friendly eyes and a cautious smile. The seemingly simple question proved confounding for the otherwise confident graduate student. Ironically, some ten years later, a version of that same question would serve as one of two primary foci for a multiyear research study we conducted to explore how teachers incorporate features of multicultural education into mathematics teaching.

In this book we detail how mathematics as one focus of our study impacted the ways teachers ultimately understood the concept of multicultural education and vice versa. For right now, however, we return briefly to the story of Shania to reveal the *seemingly simple question* that became an important backdrop for our study.

A QUESTION DURING "THE ORALS"

Shania had successfully completed the penultimate requirement for her master's degree—a daylong written comprehensive examination. It consisted of six essay-type items drawn from the content of required courses she had completed as part of the degree program. One item was devoted to multicultural education. To address it Shania defined the concept and outlined basic components of various teaching approaches.

Now with fewer than three weeks remaining in what seemed like a protracted sixteen-week semester, Shania was steadying herself for "the orals."[2] This final requirement of her degree program (completed before a three-member faculty committee) was a one-hour verbal supplement to the written examination. Although Shania felt confident about the quality of her written examination, she mentally formulated questions in anticipation of those she might be asked during the orals.

The session began smoothly and soon took on the feel of a relaxed discussion, with Shania competently fielding each question posed. It seemed she would clear this final hurdle of her degree program without a glitch until a committee member asked Shania to elaborate on the response she had submitted for the written examination item on multicultural education.

Specifically, Shania was asked, *How do we know whether or not this multicultural education stuff actually works?* Shania stammered uncharacteristically, revealing she had not anticipated *this* question.

The question dangled ominously until a different committee member[3] proffered a crisp comeback, bringing an end to the brief, albeit tense, moment, and a close to the oral examination. Shania was then asked to leave the room to allow the committee to deliberate in private about the examination outcome.

THE ORALS POSTSCRIPT: BACKDROP TO A MULTIYEAR STUDY

Fortunately for Shania, her nonresponse to the last question of her oral examination did not adversely impact the committee's evaluation of her overall performance. To be sure, she had been shaken during those final moments, yet Shania had not been the only one troubled by that fateful question. At least one committee member also had been perplexed that such a question had been presented during the high-stakes examination. Nevertheless, the significance and power of the question were undeniable.

In one fell swoop it had rendered a capable student mute and simultaneously exposed an apparent chasm between her knowledge of the theoretical scholarship surrounding multicultural education and her awareness of substantive outcomes that could be attributed directly to its implementation in classrooms. Although Shania was unprepared to offer a critical analysis of the impact of multicultural education on learning outcomes with real students, for nearly fifty years the topic itself has been central to discourses on schooling and the teaching-learning process.

As a field of study, multicultural education has offered new ways of seeing, imagining, and "doing" school. Arguably, its necessity has never been more urgent and critical to the work of educators. Figures 1.1 and 1.2 present the ethnoracial distribution of teachers and students in contemporary public schools. The dissimilarities between the ethnoracial backgrounds of students and those of their teachers highlight the continuing importance of multicultural education as well as its implications for the teaching-learning process.

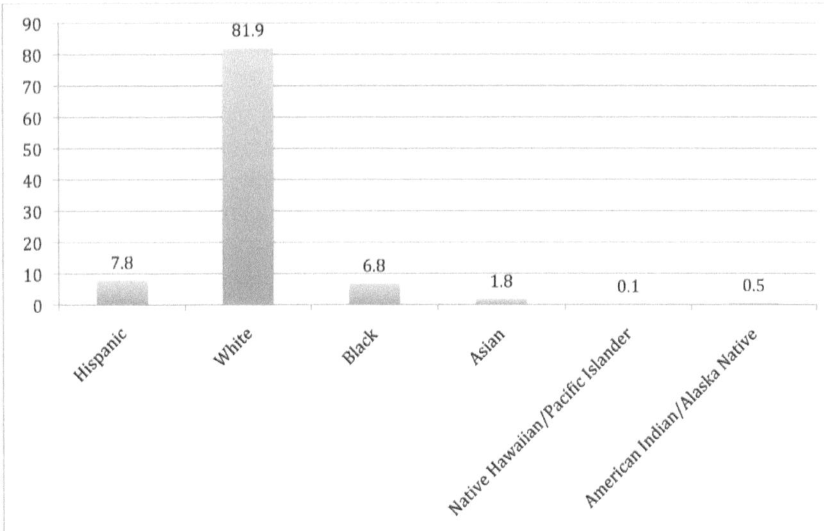

Figure 1.1 Percentage of Public School Teachers by Race/Ethnicity (2011–2012)

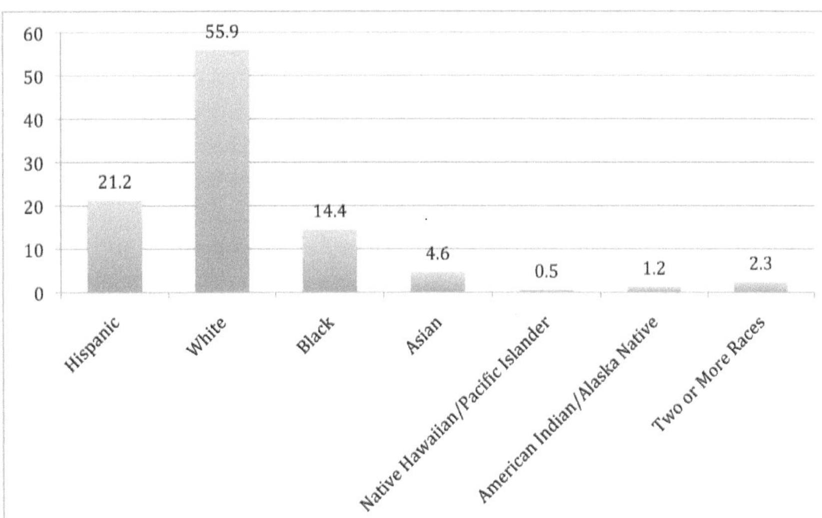

Figure 1.2 Percentage of K-12 Students by Race/Ethnicity (2011–2012)

The fact is, scholarship in the field of multicultural education began with exacting critiques of gaps in the content of school curriculum and deficits in the quality of learning experiences offered to racial minority students (Banks 2004). In time the field grew beyond the prism of race to include analyses of economic class, gender, sexuality, religion, language, and ability as critical components of culture that impact the teaching-learning process (Sleeter and Grant 2009).

Although its foci have expanded, contemporary scholars question the effectiveness with which intersections between and among these components are incorporated into multicultural teacher education in the United States (Gorski and Goodman 2011). Likewise, others have called for more deliberate efforts to demonstrate those connections in theory and research (Grant and Zwier 2011).

These issues notwithstanding, in one form or another, multicultural education continues to command the attention of various interest groups including scholars, classroom teachers, politicians, and parents of school-age children. And with its critiques of how culture and personal/group identities impact the teaching-learning process, it is not surprising that multicultural education often represents a lightning rod in discourses about schooling.

Among the questions multicultural education invariably provokes are: (1) *Does it have a place in the teaching of all curriculum content?* and (2) *What evidence is there that its inclusion enhances academic achievement?*

It may well be that persistent interest in these and other questions about the efficacy of multicultural education reflect deeper concerns about the sociopolitical contexts in which contemporary schooling occurs. For example, Kincheloe (2004) reported that schools in the United States currently exist within an atmosphere of heightened technicalization where, among other things, a premium is placed on quantitative measures of teacher and student productivity as assessed through high stakes testing. In such a climate questions about the efficiency of multicultural education in general, and its applicability to particular subject matter in particular, can be expected.

We believe careful exploration of questions such as those noted above can advance the coherence, clarity, and alignment between multicultural education scholarship and its ability to ignite teacher (and student) agency in service to higher levels of academic achievement. Conversely, we believe inattention to these questions may represent reluctance (perhaps inadvertent) to analyze the longstanding chasm between the field's widely impactful analytic/theoretical agenda on the one hand, and its technical mandate on the other.

Recognizing this conundrum, leading multicultural education theorist Geneva Gay (2002) urged researchers to engage in careful critique and analysis of those multicultural precepts most closely aligned with its real-world enactment in classrooms and schools. Warning that failure to undertake such analysis places the field at risk, she noted:

> The absence of these details limits the reformative potential of multicultural education, as well as making it susceptible to misinterpretations and distortions. Carefully executed analyses that explain the *operational* elements of multicultural education—which can be used to improve teaching at the *functional* level—will be a significant contribution to its theory and practice. (p. 46)

It seems obvious that Gay was issuing a clarion call for a new and substantively different direction in multicultural education scholarship. Rather than continuing to extol the virtues of the field's considerable conceptual foundations and frameworks, she is appealing to contemporary researchers to direct their efforts toward what is (or is not) going on with *multicultural practice* in actual classrooms with real teachers and students. We concur.

Moreover, we believe that pursuing such an agenda has the greatest potential to promote the overall growth of the field if research questions about the practice of multicultural education stimulate critique and analysis of *axioms* (or what through jaundiced eyes might be characterized as the *orthodoxy*) of the field as represented by its most revered constructs and/or theoretical foundations.

Put differently, we believe queries about the practice of multicultural education resonate most when they invite theory to fashion a stronger abutment to the real-world challenges faced by classroom teachers. Shania struggled over just such a query during her master's oral examination. What she did not know was that the *seemingly simple question* presented to her had long been a hefty challenge for multicultural education scholars. And so it was for us as we sought to study the place of multicultural education in mathematics teaching and learning.

A LOOK AT THE STATE OF STUDENT ACHIEVEMENT IN MATHEMATICS

The state of mathematics achievement in our schools is not good. In fact, underachievement in mathematics among precollegiate students represents one of the most critical educational challenges in the United States today. Some analysts offer that it jeopardizes the standing of the nation in international realms including contributions to science and engineering (Friedman 2005; Partnership for 21st Century Skills & Board 2006). Others propose that it puts many young people at risk of being ill prepared to negotiate life effectively in our increasingly high-tech society (President's Council of Advisors on Science and Technology 2010).

Concerns such as these may be especially warranted when coupled with data from mathematics assessments. Results from the most recent National

Assessment of Educational Progress (NAEP), for example, reveal that although average mathematics scores for fourth graders increased by one point from the 2011 assessment,[4] elementary students across the United States struggle with concepts such as measurement, place value, and operations (Institute of Education Sciences 2011).

On the 2011 assessment, only 14 percent of fourth graders answered correctly a problem requiring them to identify which was larger, the result of a given number divided by 18 or the result of that same number divided by 15.[5]

By design, large-scale assessments such as the NAEP generate information about the collective performance of students in schools across the nation. Yet contemporary schools are populated with children and adolescents from a wide array of cultural, ethnoracial, and linguistic backgrounds. Likewise, they come from myriad socioeconomic circumstances. Therefore, when data generated by the NAEP has been disaggregated, many of its confounding facets are brought to full relief.

Consistently, the data reveal that children and adolescents most affected by the mathematics underachievement epidemic are those from economically poor and African American, Latina/o, and Native American backgrounds. Among other things, this means that many future career options that depend on solid knowledge of mathematics beyond the most rudimentary levels will be foreclosed to most of these students.

Moses and Cobb (2001) drew parallels between this reality and 1960s struggles nationwide for racial equality and social justice. They called for mathematics underachievement to be treated as a contemporary civil rights issue. This is because despite the slight improvement in average fourth grade mathematics scores since the last administration of the NAEP, a breakdown of those scores by race and ethnicity shows that there is still a significant difference between the achievement of ethnoracial minority students and that of their White peers. This difference has increased in recent years (Institute of Education Sciences 2011) and researchers report that evidence of its presence surfaces well before fourth grade.

In a review of the research on early childhood mathematics, Clements and Sarama (2007) detailed nuances in the early onset of mathematics underachievement. They reported that students from ethnoracial minority backgrounds, even as they start formal schooling, demonstrate significantly lower levels of achievement when compared to White students with respect to numerical, arithmetic, spatial/geometric, patterning, and measurement knowledge.[6] In short, some children enter kindergarten needing to "catch up" to their peers in terms of their understanding of mathematics ideas (Clements and Sarama 2007).

Such findings are a clear indication that a "gap" exists between the achievements in mathematics for children from different cultural/ethnoracial backgrounds. The reason for the gap is multifaceted. Scholars agree that it is fueled and perpetuated by long-standing inequities, including gross variations in the quality of schooling available to students from different backgrounds (Darling-Hammond and Sykes 2003).

Also implicated is the fact that many students simply do not have sufficient opportunities to learn (Hilliard 2003; Martin 2000) high-quality mathematics. The latter problem was disturbingly illustrated in findings from a recent statistical analysis of data drawn from the Early Childhood Longitudinal Study—Kindergarten (ECLS-K) national database.

A team of researchers calculated odds ratio (OR) to examine the likelihood of particular groups of elementary-level students being afforded the chance to study higher-level mathematics by eighth grade (Faulkner et al. 2014).[7] To control for the reverberating effects of early underachievement in mathematics, the researchers used data collected only from those children who had demonstrated *high mathematics achievement* at the fifth grade based on standardized test scores.

The chilling findings revealed that "a student who is Black has an OR of 0.68, meaning an odds placement that is approximately two-thirds that of his or her White peers" (p. 302). Moreover, gaining access to study of algebra for Black students was more greatly affected by recommendations from their teachers than by the actual demonstrated abilities of the students. The researchers noted,

> When it comes to Black students, performance data appear to have a reduced influence on placement [in Algebra] and teacher evaluation an enhanced influence. That is to say, even in the light of objective measures of achievement and potential, the demonstrated mathematics acumen of high-performing Black students cannot be counted on (as it does for their White peers) to differentiate them academically from lower performing same-race peers. In short, Black students confront an untenable impediment in that their Blackness (or . . . the teachers' implicit responses to these students' Blackness) serves as an invisible, albeit formidable obstacle to gaining access to higher level mathematics courses, irrespective of their demonstrated performance. (p. 306)

Thus, it would appear that while findings from large-scale mathematics assessments like ECLS-K and NAEP provide important information about the state of student achievement, at the same time they expose entrenched disparities in the protocols and conventions at work in many of our nation's schools. We believe a particularly troubling disparity surrounds the thorny issue of the quality of instruction.

ISSUES SURROUNDING HIGH-QUALITY MATHEMATICS TEACHING AND LEARNING

Throughout the world, the formal schooling made available to young children foretells to a great extent the likely quality of their adult lives. Widespread consensus holds that a standard element of the best schooling for all students should be high-quality mathematics *teaching and learning*.

In the classroom environment, the latter (learning) is greatly affected by the former (teaching). Therefore, it is not surprising that in the United States this inextricable relationship has been emphasized in numerous reports including the *National Commission on Teaching and America's Future* (2003); *No Dream Denied, The Mathematics Education of Teachers Report* (CBMS, 2000); and *Mathematical Proficiency for All Students* (RAND Mathematics Study Panel, 2003). All have shown that what counts most in facilitating and improving student knowledge of and achievement in mathematics is the quality of the instruction received.

Simply put, **teachers matter**. And when it comes to promoting deep understanding and high achievement in mathematics, teachers matter in the most critical and most fundamental ways. However, findings from national and international assessments like the NAEP, Program for International Student Assessment (PISA), and ECLS-K suggest that for many teachers, it is exceedingly difficult to promote the desired outcome of deep mathematics understanding for all their students.

In part the difficulty rests on the fact that much mathematics underachievement is traceable to ill-formed understandings of fundamental mathematics ideas. Most children have a multitude of encounters and experiences with mathematics in their lives outside school. Even so, it is doubtful that many start formal schooling with a negative attitude toward mathematics.

Rather, a more likely scenario is that before they begin formal schooling, and early on in their schooling careers, many children acquire cognitive templates on which they form incomplete or erroneous understandings about mathematics (e.g., Baroody and Wilkins 1999; Clements and Sarama 2007; Ginsburg, Lee, and Boyd 2008; Griffin, Case, and Siegler 1994; Hiebert 1986). Then, in the context of formal study of "school" mathematics, children invariably draw upon their experiences and emerging understandings.

Unwittingly they further embed malformed templates (or acquire new ones) as they move through the grades. For their part, teachers discover that before they can help these children advance in their mathematics understanding, they must help them "unlearn" distorted or dysfunctional ways of thinking about basic concepts.

Merging such an undertaking with the cumulative character of mathematics itself creates an intricate web of pedagogical challenge for even the most confident teachers. The reality, however, is that some teachers are not "up" for this challenge and there are several issues that may contribute to this unfortunate situation.

In many school districts, teachers feel overwhelmed due to an ever-increasing volume of both instructional and noninstructional tasks that are part of the job. A common lament is that there isn't enough time in the day for actual teaching. Correspondingly, it is not uncommon for some teachers to hold the belief that some students are naturally predisposed to learn mathematics while others are not.

Whether these particular issues operate independently or in tandem, the ultimate losers are children who miss out on high-quality mathematics instruction. Moreover, for a child the reverberation of such an educational loss is greatest when it is initiated in the primary grades. This is because teachers at the primary level have a heavy professional assignment in that they must provide the *foundation* for children's mathematics studies in later grades.

Ironically, it happens that those who teach in the earliest grades, and arguably have the heftiest pedagogical challenge, often do not possess the mathematical knowledge or reasoning skills needed to teach mathematics in a way that results in children's acquisition of deep and transferable understandings (RAND 2003; Wilson, Flodin, and Fernini-Mundy 2001).

Hence, task intensification (Apple and Jungck 1990) and the resulting limitations on instructional time notwithstanding, without substantive professional intervention for such teachers, the mathematics underachievement that surfaces early in children's school careers is unlikely to dissipate. It may even increase as these children progress through the grades.

In other words, where a teacher's own mathematical knowledge is slight, her ability to promote deep mathematical understanding among her students is not only compromised, but it also may even be nonexistent.

Researchers who focus on the knowledge needed to teach mathematics have reported that improved student achievement depends greatly on the professional development of teachers (Borko 2004; Hill and Ball 2004; Hill, Rowan, and Ball 2005; Sowder 2007). At least two central questions are begged by the crisis of mathematics underachievement, its early onset, and teachers' abilities to help change the tide. These are:

1. How does the mathematics pedagogy of primary grade teachers, along with their own professional identities as mathematics teacher-learners, facilitate (or constrain) children's opportunities to acquire deep understanding of early mathematics ideas?
2. How do teachers make use of children's outside-of-school mathematics experiences to facilitate learning of school mathematics?

We explored these and other questions in our multiyear study called Nurturing Mathematics Dreamkeepers.

TOWARD BRIDGING CHASMS

Nurturing Mathematics Dreamkeepers[8] (NMD) was a quasi-experimental research study of how teachers incorporate elements of multiculturalism into early mathematics instruction. Its main goal was to facilitate teachers' abilities to structure instruction in such a way as to promote deep understanding of fundamental mathematics concepts among young students.

Teachers in the study were required to participate in extensive professional development designed to enhance their content and pedagogical content knowledge of mathematics (Shulman 1986). Content knowledge represented those concepts, principles, algorithms, and so on that constitute standard components of the elementary mathematics curriculum for the school district in which the study occurred.

Pedagogical content knowledge comprised "the most useful forms of representation of those ideas, the most powerful analogies, illustrations, examples, explanations, and demonstrations—in a word, the ways of representing and formulating the subject that make it comprehensible to others" (Shulman 1986, p. 9).

An especially critical task for us as researchers was to sensitize teachers to the realization that the "analogies, illustrations, examples . . ." they draw upon to help their students understand mathematics reflect worldviews and sets of assumptions (most likely those that align with a given teacher's own experiences as a learner of mathematics) about the subject. Essentially, we challenged teachers to consider that the particular worldview or filter through which mathematics (or any subject for that matter) is taught typically remains invisible to a teacher unless and until she becomes aware of and sensitive to its presence.

And so NMD provided a forum for the teachers to analyze how such a filter might be informing their own instructional practices. It permitted them to recognize manners in which biases and taken-for-granted axioms embedded in their own teaching, if unmonitored, could present advantages for some students while simultaneously erecting barriers and disadvantages for others.

Teachers were challenged to become aware of, and if necessary make adaptations to, their instructional frameworks in order to better align them with the experiences and the diverse learning needs of their students. Thus, beyond its goal to enhance teachers' content and pedagogical content knowledge of mathematics, NMD promoted awareness and understanding among the teachers of ways culture impacts the teaching-learning process.

Ultimately, we sought to advance an orientation to instruction that facilitated teachers' abilities to recognize and teach toward the nuances of cultural experience as manifested in their students' outside-of-school experiences with mathematics. The two complementary frameworks for the NMD study were **standards-based mathematics teaching** and **culturally relevant pedagogy**.

What Is Standards-Based Mathematics Teaching?

Standards-based mathematics (SBM) teaching refers to practices consistent with goals set forth by the National Council of Teachers of Mathematics (NCTM) in the document *Principles & Standards for School Mathematics* (NCTM 2000). The document states that teaching "requires understanding what students know and need to learn and then challenging them and supporting them to learn it well" (p. 16).

In short, the standards are a call for teaching mathematics with a goal of promoting deep understanding for *all* students. While we added the emphasis in the previous sentence, NCTM also emphasized the importance of providing opportunities to develop understanding for "all students" by the inclusion of the *Equity Principle*. The Equity Principle states that teachers must include "high expectations and strong support for all students" (p. 12).

With the *Equity Principle*, standards-based teaching characterizes the classroom practices and pedagogical stances of teachers who operate from the assumption that a student's mathematical reality is not independent of that student's ways of knowing and acting outside of mathematical contexts. This implies that what a student sees, understands, and learns is constrained and supported by what that student already knows. It also suggests mathematical learning is a process of transforming one's knowing and ways of acting (Simon et al. 2000, p. 584).

Franke, Kazemi, and Battey (2007) note that support for such learning "comes from *knowing the students* [emphasis added], the situation, and the content and then making decisions that support interaction that productively engages students in moving their ideas forward" (p. 228). It's important to emphasize that the *Equity Principle* is not an "add-on" to the NCTM Principles and Standards, rather it is central to all pedagogical moves (Hiebert et al. 1997). The Equity Principle has been reiterated in a recent publication, *Principles to Actions: Ensuring Mathematical Success for All* (NCTM 2014).

Practice that is consistent with standards-based goals and principles provides equitable opportunities for all students in all aspects of classroom interactions. It requires that teachers convey a fundamental respect for individual students, what they bring to the classroom in terms of knowledge and experience, and what their individual contributions can add to their peers' mathematical understanding (Hiebert et al. 1997; NCTM 2014).

In mathematics, the ability to explain and apply more complex concepts and operations builds directly upon the depth of understanding one acquires of its more fundamental ideas and principles. This structure of the discipline was critical in our selection of primary grades for our study.

The focus made it especially important that we design the professional development component in such a way as to help teachers recognize limitations of perceiving *school* as the only (or even the most consequential) site of students' engagement with mathematics. This is because at the primary grade levels, children have spent more of their lives outside rather than inside schools.

We wanted teachers to recognize and appreciate that informal experiences with mathematics, as embedded in children's everyday cultural interactions, are a crucial component of enhancing mathematics teaching and promoting deep understanding.

Three Tenets of Culturally Relevant Pedagogy (CRP)

A central idea undergirding NMD was that in order to enact SBM instructional practices, teachers need to understand culture as a basic component of the teaching-learning process. We defined culture as "the consistent ways in which people experience, interpret, and respond to the world around them" (Marshall 2002, p. 8) including the tendencies or orientations commonly exhibited in academic contexts by students from the same cultural/ethnoracial community.

Scholars identify culture as fundamental to the broader schooling enterprise and they have called upon educators to adopt critical professional orientations (Gay 2002; Giroux 2005; hooks 1994; Kincheloe 2004), such as *CRP* (Ladson-Billings 1994), which reflects understanding and appreciation of its impact.

According to Ladson-Billings (1994), there are three primary tenets or components of CRP: high academic achievement, cultural competency, and sociopolitical consciousness. Teachers who effectively promote *high academic achievement*, for example, must use an array of resources and presentation formats that align with their students' learning styles or learning tendencies. Tailored to the needs and experiences of the particular students, cultural relevance as manifested in culturally responsive teaching practice (Gay 2002) in one classroom may look quite different from that in another classroom.

Even so, all teachers whose pedagogy exhibits cultural relevance find ways to incorporate students' outside-of-school cultural realities into instruction, thereby enhancing students' interest in and success with academic content. Hence, instruction that is grounded in cultural relevance is "based on the assumption that when academic knowledge and skills are situated within the lived experiences and frames of reference of students, they are more personally meaningful, have greater interest appeal, and are learned more easily and thoroughly" (Gay 2002, p. 106).

The second component of CRP, *cultural competency*, represents teachers' abilities to "capitalize on the cultural practices and sensibilities of their students" (Nasir, Hand, and Taylor 2008, p. 219). At the same time, however, it represents teachers' introspective capacities to acknowledge themselves as cultural beings.

Teachers who embrace CRP recognize that without constant vigilance (via reflection on the personal subjectivities they introduce unwittingly to the teaching-learning process), it is possible to inadvertently squelch the sense of agency most students naturally exhibit in their early school careers. In so doing, teachers can diminish academic learning for different students through their (the teachers') own actions.

Finally, *sociopolitical consciousness*, the third component, relates to teachers' orientations toward status, resource, and power differentials among diverse groups in society, and the impact these inequities have on students in schools. Teachers who adopt CRP recognize that intersections between race, class, and gender, for example, more commonly than not, still do matter in who gains access to high-quality schooling (Bryk et al. 2010; Noguera 2001; Orfield and Lee 2006).

As such, these teachers strive to create and maintain equity in their classrooms through building coalitions with other teachers, parents, and students themselves that help counteract the negative impact of disparities and inequities in the larger society, and the resultant manifestations in schools.

Merging Multicultural Education and Early Mathematics

In a comprehensive description of the history and development of multicultural education as an area of academic scholarship, Banks (2004) detailed its five dimensions: content integration, knowledge construction, prejudice reduction, equity pedagogy, and empowering the school culture and social structure. The relevance and necessity of attending to the diversity of culture in contemporary schools and classrooms is embedded in each dimension.

Still, it was the idea of *equity in pedagogy* that was most fundamental to the NMD study. Although represented in the study's two complementary frameworks (CRP and SBM), it was most explicitly reinforced through the notion of CRP. Marshall (2002) detailed connotative and denotative meanings associated with the concept of pedagogy as part of multicultural education. She noted,

> Pedagogy refers to the overall professional orientation teachers bring to their work in schools. Teachers' pedagogy is reflected in their perceptions of the teaching role and their assumptions about students. It is also accented in the instructional strategies teachers employ and the learning atmospheres they create in classrooms. As a dimension of multicultural education, the concept of equity pedagogy is based on the premise that all students can learn. Consequently, teachers who embrace this premise work to create learning atmospheres that demonstrably

> heighten the likelihood that all students will learn. Basic to the concept . . . is the idea that students learn in diverse ways. These diverse ways or styles are greatly influenced (but not dictated) by the personal experiences, worldviews, and cultural backgrounds students bring to the learning environment. . . . [No] particular style of learning is subordinate (or superior) to another. Teachers . . . incorporate diverse teaching techniques and methods in an effort to facilitate the academic achievement of all students. An equity pedagogy requires teachers to integrate diverse content into the standard curriculum. In fact, it may even require that they replace the standard curriculum with content that is more relevant to the academic needs, personal experiences, and cultural orientations of their students. (p. 25)

Teacher education scholars and researchers have long advocated for alternative critical pedagogies that have the potential to deliver to contemporary children the "promise" of high-quality education. In particular, cultural relevance has been widely embraced by mathematics education scholars (Gutstein 2003; Leonard et al. 2010; Nasir, Hand, and Taylor 2008) as a critical pedagogy of possibility.

We believe the appeal of CRP, and by extension the need for multicultural education, is in part owing to the enormity of the situation facing the society at large vis-à-vis underachievement, low-quality schooling, and ultimately lost opportunity. Prior to the NMD project, however, scholars had not attempted to describe and analyze the more vexing and troublesome challenges that accompany attempts to incorporate the triadic focus of CRP into the *real-world lives* of mathematics teachers at the *primary level*—the most critical stage of mathematics teaching and learning.

BEYOND AXIOMS AND ORTHODOXY

In this book we turn a critical lens on CRP with the aim of offering an assessment of its robustness as a framework for professional development for practicing K-2 teachers. Through detailing our multiyear research study, we have unpacked some of the real-world complexities of promoting the Equity Principle in mathematics teaching vis-à-vis CRP.

We describe and critique the hands-on versus minds-on quandary that confronted K-2 teachers in our study as they worked to facilitate deep mathematical understandings among primary grade students. Additionally, we explain how the teachers made sense of the reality of culture in the teaching-learning process particularly as related to mathematics. Using case studies we detail how teachers meshed their own culturally influenced ways of being (along with their identities as primary-level teachers) with the mandate to provide high-quality, SBM instruction.

In this chapter we have discussed the merger between multicultural education (CRP) and high-quality mathematics (SBM) that served as the frameworks for the study. The remaining chapters present a full discussion of the NMD project, including the extensive professional development component, case studies of participants, and its implications for future study.

Specifically, in chapter 2, *Beyond Licensure Renewal Credits in Teacher Professional Development*, we provide an in-depth description of features of the study. This includes the structure of the professional development component, participant recruitment including the year-long, repeating cohort structure, and data collection protocols.

Chapter 3, *Math Learners and Cultural Beings: K-2 Teachers As Dreamkeepers,* introduces nine teachers who will serve as the focus of the book and represent the primary and secondary cases drawn from the larger study. This chapter details the setting for NMD including background information about the county and school district where the study took place. Also we describe each school's leadership team along with the climate/context in which each teacher worked.

Then, in chapter 4, *Dreaming a Bigger Dream: The Primary Case Teachers*, we discuss four teachers' journeys to take on the challenge presented by the NMD project. That is, we detail each teacher's struggles and successes to enhance the depth of her own mathematics content knowledge; beliefs and dispositions toward cultural diversity; and her skills and ability to merge CRP and SBM to facilitate academic success in mathematics for young students.

Chapter 5, *A Theory/Practice Chasm: The Secondary Case Teachers*, takes a look at the experiences of the five teachers who struggled in unique ways to merge the project's two foci. In examining their collective experience, one of the undergirding issues of this chapter explores whether critical multiculturalism represents the proverbial "moving target" (when presented in tandem with mathematics) as a professional objective for some teachers. Using illustrations from our secondary case study teachers, we reveal how our efforts to merge CRP and SBM in professional development, in some cases, brought to full relief the enormity of the challenge.

Finally, chapter 6, *Toward a Liberating Professional Identity*, we offer our reflective analysis and critique of the NMD project as a whole. Therein we explore our concept of *planned affirmative intersectionality* as a component of teacher professional development. We conclude by using NMD findings to theorize a revisioning of the notion of intersection as part of what we call *a liberating pedagogical identity.*

EPILOGUE

The NMD study was an exploration of what happens when multicultural education is merged with the mandate for high-quality mathematics teaching and learning. Put differently, it was an effort to determine the extent to which "the stuff actually works" in the messy, real-world context of classrooms. In hindsight we recognize that it was also a challenge not dissimilar to the one presented to Shania, the graduate student introduced at the start of this chapter. Not unlike Shania, we too were humbled by its enormity. We wrote this book to tell the tale.

NOTES

1. Except where we specifically refer to ourselves, all other proper names of persons, schools, and locations throughout this book are pseudonyms.
2. Eventually the orals were dropped as a component of the master's degree when the Department of Public Instruction issued a mandate for all teacher education programs in the state to be revised. At our institution, the two-part comprehensive examination was discontinued and replaced with *authentic* end-of-degree program assessments such as portfolios and classroom-based action research projects that included self-studies.
3. Lead author Pat Marshall served as Shania's academic adviser and chairperson of her oral examination committee. Witnessing her discomfort, Marshall reflexively "rescued" Shania by turning to the colleague and replying, *well, we don't know if it works since it's never really been tried in these schools*. The colleague nodded in agreement suggesting that Marshall's response had sufficed and that it had represented a pithy assessment of the tepid reception multicultural education was receiving in the local school district.
4. The NAEP is administered every two years, and fourth grade is the earliest level at which it is administered. Scores reported in 2011 revealed a twenty-eight-point increase from the 1990 assessment.
5. The released items from 2013 include only those in which more than 50 percent of students surveyed answered correctly.
6. Examples of numerical knowledge include the ability to identify a one-to-one correspondence, count on, and count by twos or fives. Arithmetic skills include explaining sums, differences, products, and quotients as well as recognizing when each is an appropriate operation to use. Spatial/geometric skills include measurement, characterizing two- and three-dimensional figures, and connecting geometric representations to arithmetic operations.
7. Higher-level mathematics was identified as study of eighth-grade algebra. Study of algebra is widely recognized as the "gateway subject" to gaining access to more advanced mathematics study such as calculus and differential equations. Knowledge of these and other areas of mathematics is typically required to gain access to particular college majors and professional fields including engineering, architecture, and various physical science (physics) and applied mathematics (statistics) fields.
8. The project was funded by a grant from the United States National Science Foundation (award #035341). Any opinions, findings, and conclusions noted in this book are those of the authors and do not necessarily reflect the views of the National Science Foundation.

Chapter 2

Beyond Licensure Renewal Credits in Teacher Professional Development

In the previous chapter, we introduced NMD, a multiyear research study that merged multiculturalism and mathematics in an exploration of how primary grade teachers integrate components of CRP and SBM teaching.

An aim of NMD was to try to provoke curiosity about the possibilities for pedagogical experimentation. This we did by providing teachers a professional development experience that was substantively different from any they had participated in previously.

Our hope was for the teachers to acquire critical understanding of the impact of culture on the teaching-learning process. Correspondingly, we wanted them to contemplate the implications of mathematics underachievement among students of color as it occurs in the earliest grades.

The road to the successful completion of NMD was not smooth. Rather, it was decidedly bumpy, with numerous detours along the way. Therefore, in this chapter we discuss the mechanics of the study with a focus on some of its most elemental features.

We begin with a discussion of how we came up with the idea for NMD, followed by a description of our research design. Next we examine complexities we encountered in conducting our longitudinal study in the public school context. Among the issues we explore are the difficulties surrounding our plans for participant recruitment, delivery of the professional development intervention, and implementation of the multipart data collection protocol. We conclude this chapter with an overview of the evolution of the NMD research team.

A NATIONAL CALL FOR HIGHLY QUALIFIED TEACHERS

The idea for NMD was a response to *No Child Left Behind* (NCLB),[1] the 2001 federal legislation in the United States that authorized massive school reform. Ostensibly, the purpose of NCLB was to guarantee that every child who attends the nation's public schools receives a high-quality education. With such an unassailable objective it was not surprising that NCLB received bipartisan support in the Congress.

The primary mandate of NCLB was for precollegiate educational experiences to be enhanced to such a degree that upon school completion, every child would be prepared to take advantage of a broad range of opportunities. An especially noteworthy feature of NCLB was its explicit focus on the academic progress of students from targeted populations. These included ethnoracial minorities, students from poor and working-class economic circumstances, English language learners, and students with special education designations.

At the time of the legislation's passage, thousands of students from the targeted population groups were languishing in schools from which they seemed destined to exit with minimal skills and limited life choices. In a speech delivered at the 91st Annual Convention of the National Association for the Advancement of Colored People, the nation's oldest civil rights advocacy organization, President George W. Bush characterized the schooling experiences of many poor and ethnoracial minority children as "the soft bigotry of low expectations" (eMediaMillWorks 2000).

Widespread underachievement and school failure among children and youth from what NCLB designated as targeted populations had long been a scourge on public education. Yet evidence of the problem typically had been buried in academic achievement statistics reported as aggregated data. Whether by happenstance or design, school district officials throughout the nation were routinely concealing, and in many cases perpetuating, the dismal performance outcomes of many children and youth from targeted populations.

To end those practices, the NCLB legislation mandated that testing data be disaggregated in a manner that would show the academic yearly progress (or lack thereof) of each targeted group relative to the entire population of students within a given school or district. Furthermore, sanctions were to be imposed on those districts that did not demonstrate growth for targeted populations. Thus, with the passage of NCLB it seemed the protracted reign of academic underachievement and failure for "other people's children" (Delpit 2012) was finally receiving the national attention it merited but had long been denied.

Despite its laudable goals, however, various features of the new law came under immediate attack. Testing and strict monitoring of school performance goals were among its more contentious features.[2] Lacking financial support from the federal government, most districts found it virtually impossible to adhere to the spirit, let alone the letter, of this massive legislation. But these issues notwithstanding, the one aspect of NCLB that even its severest critics could not assail was its call for an upgrade of the nation's teaching force.

The focus on teachers in the NCLB legislation was an especially urgent issue in schools with large numbers of students from the targeted populations. This was because indisputably, it was those children who for decades had been left behind. Scholars report that schools serving large percentages of ethnoracial minorities and/or large numbers of students from economically poor families are often staffed by significant percentages of inexperienced teachers and/or those assigned to teach subjects for which they have minimal professional preparation (Darling-Hammond 2004).

To combat this, the NCLB law called for every teacher to possess a depth of knowledge in relation to the content they were required to teach. In the terminology of the law, every teacher was to be "highly qualified." But since the law had made specific mention of targeted populations, being highly qualified was not limited to teachers' content or subject matter knowledge. Instead, it extended to skills and dispositions to teach students from an array of cultural backgrounds and circumstances. Thus the idea for NMD was an effort to address the NCLB mandate that called for every child to be taught by a highly qualified teacher.

Our plan was to enhance the knowledge, skills, and dispositions of primary grade teachers to teach mathematics in a manner that would be both relevant and responsive to the culturally diverse backgrounds and experiences of kindergarten, first-, and second-grade children in one large urban school district.

There were three core principles of the NMD project: (1) mathematics understanding is critical to function effectively in a highly technological society; (2) the single most critical factor in mathematics underachievement is the gap between the culture of school and the culture of home; and (3) research in classrooms indicates that mathematical understanding can be nurtured in all students.

MAKING GOOD ON THE LAW

The highly qualified teachers' mandate of NCLB was both enticing and challenging. As co-principal investigators for the NMD study, we were intrigued by its complexity and its paradox. The law called for persons (teachers), who had been educated and were currently working in school systems throughout the nation, to make unprecedented changes in those very systems (Ball and Cohen 1999).

Aware of this inherent conflict in the mandate, we grounded NMD on four assumptions. First, teachers are not likely to become change agents unless they are engaged in an appropriate intervention that facilitates and supports their ability and efforts to make change.

Second, any intervention designed to facilitate teachers becoming change agents in schools must be structured in such a manner that teachers develop knowledge, skills, and dispositions that are rooted in their classroom practices. In that way they can choose and adjust learning activities in a manner that is likely best suited to and proper for teaching particular concepts to particular students. The intervention itself, however, must be designed in such a way that it allows teachers to gradually take ownership of, sustain, and perhaps even promote the desired change.

Third, the intervention needs to take place during the earliest years of formal schooling in order to set a firm learning foundation for children. Research indicates that poor teaching has a lasting negative impact. Sanders and Horn (1998), for example, reported that youngsters taught by an incompetent teacher one year continue to suffer effects of that poor teaching even when in subsequent years their teachers are more competent.

Finally, our fourth assumption was that when it comes to enhancing mathematics learning for children from targeted groups, the success of an intervention can be best enhanced if it draws on and coordinates the scholarship of early mathematics teaching with that which addresses the impact of culture on the teaching-learning process. Such coordination is important because neither mathematics nor the teaching-learning process itself is culture-free or value neutral (Ladson-Billings 1997; Murtadha-Watts and D'Ambrosio 1997). Rather, children draw upon what they already know (e.g., information they acquire through socialization received in homes and communities) to make sense of what they are taught in schools.

Indeed, a plethora of scholarship indicates that teachers who are most successful with students from targeted groups are those who have developed facility in drawing upon the learners' outside-of-school experiences to promote understanding of school content (Boykin et al. 1997; Cabrera et al. 2014; Deyhle 1983; Irvine 1991; Ladson-Billings 1994; Lipka 1991, 2005; Moll et al. 1992; Nasir, Hand, and Taylor 2008).

At its core, NMD was a project about teachers' professional development. As detailed in the previous chapter, our focus was steadied on how primary-level teachers understand and adopt standards-based teaching practices (NCTM 2000) that promote children's conceptual or deep understanding of the big ideas in early mathematics.

Additionally, NMD was an exploration of how teachers understand the impact of culture on the teaching-learning process, and how they adapt components of CRP (grounded in knowledge of their own and their students' cultural ways of being, worldviews, and tendencies toward learning) in their mathematics instruction.

Through these foci we aimed to facilitate an epistemological shift (Simon et al. 2004) in the pedagogical orientations of the teachers, and to provoke changes in their instructional behaviors and professional worldviews. We theorized that the degree to which such a shift did occur would be attributable, in part, to the teachers' own enhanced understanding of mathematics. In turn, their enhanced understanding, we reasoned, would result from active engagement in reflective analysis of their teaching (actual lessons) and evaluation of their students' mathematical understandings.

For us, the teachers themselves needed to perceive a need to alter (or change outright) their pedagogy. Such a change would necessarily involve replacing practices characterized by telling and demonstrating procedures with a pedagogy in which teachers thoughtfully (a) teased out and affirmed children's out-of-school mathematics experiences; (b) recognized the form and function of children's emerging understandings of mathematics "domain" knowledge (Nasir, Hand, and Taylor 2008); and (c) orchestrated learning moments (informed by the children's own learning goals) that promote deep mathematical understanding.

A critical issue in improving early mathematics instruction, however, is the pedagogical knowledge and general comfort teachers demonstrate with the content itself. We wanted teachers to promote children's deep understanding in mathematics. At the same time, we recognized that in order to do this, many primary-level teachers would need to enhance their own content knowledge and skills, and adopt different dispositions toward the teaching of mathematics (Ball 2000; Borko 2004; Ma 1999).

This meant that the teachers in our study would need to revisit their own mathematics stories (McCulloch et al. 2013). In so doing, they would have to find the professional (as well as the personal) daring to reflect upon their own histories and experiences as mathematics learners. Perhaps the teachers would even have to acknowledge that they themselves needed help in understanding the mathematics they were required to teach as part of the curriculum.

MODIFYING AND REVISING A LONGITUDINAL RESEARCH PLAN

By most indicators of research involving practicing classroom teachers, the agenda for NMD was ambitious. It was longitudinal, and not unlike other multiyear studies, NMD encountered glitches and problems that necessitated the reformulation of an otherwise sound research plan. Once we began the study we attempted to retain as many of our original ideas as possible, yet the actual conduct of NMD required both slight and substantial changes throughout its duration.

Many of the changes were necessitated by the dynamics of the public school context in which the study took place. Two broad categories of change that warrant brief explanation are *modifications* and *revisions*. The former represented the "nipping and tucking" activities that occurred once we put the ideas-on-paper into action.

Most modifications emerged organically thereby resulting in only minimal disruption to the overall flow of the study or the work demands for any one member of the research team. Revisions, on the other hand, typically resulted from an event or situation that threatened the integrity, progress, and/or stability of the project itself. These changes were often unexpected and indeed, a series of revisions was set off when a tragic event compromised team cohesion and esprit de corps.

In hindsight, the degree of modification and revision required to conduct NMD was somewhat ironic because the research plan itself had been exceptionally well vetted. The pre-funded proposal underwent a rigorous, multistage evaluation to determine its soundness. Similarly, an advisory board of nationally recognized experts was organized and consulted periodically throughout the study to assure that the work of the team remained on course. Even so, once underway NMD endured various challenges.

A goal of this book is to discuss NMD in a manner that provides useful insights to others interested in classroom-based research. Therefore, perhaps different from most published accounts of multicultural education and mathematics teaching and learning studies, we share some of the pitfalls we encountered and the resultant changes we made during the span of NMD. In the next section we describe the research design, research questions and hypotheses, and data sources.

A MULTILAYERED RESEARCH DESIGN

The NMD project began with a complex research plan. It was organized as an embedded experimental mixed methods design wherein the *qualitative* data (e.g., interviews) was used to support the *quantitative* data (e.g., surveys, questionnaires) (Creswell and Plano Clark 2011). In this design, the qualitative data was embedded within the quantitative components. Although planned with these specifications, several glitches caused us to have to make changes in the design.

One of the more significant unanticipated changes was that our plan to include a participant control group could not be implemented until the study's final data collection year. (The reasons for this deviation from a traditional quasi-experimental design are explained in subsequent sections of this chapter.)

Our inquiry was guided by the following three central research questions and six hypotheses:

1. How do K-2 teachers reorganize their teaching toward a knowing-in-practice that is grounded in conception-based, CRP in terms of: articulating student mathematical conceptions; deciphering the intended mathematics of standard K-2 mathematics curriculum; and hypothesizing about paths from extant to intended conceptions and selecting and/or adjusting teaching-learning tasks to promote such advancement in their students?
   H_1 Dreamkeepers will develop knowing-in-practice that is more consistent with the project intervention than standard practice teachers.
   H_2 Cohort I Dreamkeepers will develop knowing-in-practice more consistent with the intervention pedagogy than Cohort II and Cohort III.
   H_3 Cohort II Dreamkeepers will develop knowing-in-practice more consistent with the intervention pedagogy than Cohort III.
2. How do teachers reorganize their mathematics instructional practice to reflect intercultural sensitivity and to incorporate cultural relevance?
3. To what extent do teachers' evolving understandings and practices lessen the achievement gap in mathematics between African American and White American K-2 students?
   H_4 African American and White American students of Dreamkeepers will outperform their counterparts of standard practice.
   H_5 African American and White American students of Cohort I Dreamkeepers will outperform their counterparts of Cohort II and Cohort III Dreamkeepers.
   H_6 African American and White American students of Cohort II Dreamkeepers will outperform their counterparts of Cohort III Dreamkeepers.

In order to address our research questions and hypotheses, we used a variety of data sources generated primarily from our teacher participants, but also drawn from school leaders and students. Over ten different sources of data were collected with instruments created or adapted for NMD. Table 2.1 lists all the major NMD data sources including information about type (quantitative, qualitative, both); frequency of collection; and the specific NMD data year(s) in which it was collected.

We analyzed our data using a variety of both quantitative (e.g., ANOVA, Exploratory Factor Analysis) and qualitative (e.g., thematic content analysis) approaches. Also, we produced instruments to assist with our analyses.

Table 2.1 NMD Data Sources

Source	Data Type	Collection Frequency	Data Collection Year
E-Forum Asynchronous Discussions	Qualitative	Monthly	III
Field Notes of Videotaped Lessons	Qualitative	Four data sets	I, II, III
Mathematics Survey (Teachers)	Quantitative	Semiannually	I, II, III
Mathematics Survey (first and second graders)	Qualitative and Quantitative	Semiannually	III (Cohort 3 only)
Mathematics Interviews (Kindergarteners)	Qualitative and Quantitative	Semiannually	III (Cohort 3 only)
Multicultural Teaching Concerns Survey	Quantitative	Semiannually	II, III
One-on-One Interviews (Teachers)	Qualitative	Semiannually	I, II, III
One-on-One Interviews (School Leaders)	Qualitative	Semiannually	II, III
Project Artifacts	Qualitative	Per retreat	I, II, III
Videotaped Mathematics Lessons	Qualitative and Quantitative	Four data sets	I, II, III
Videotaped Lesson Guided Reflections	Qualitative and Quantitative	Four data sets	I, II, III

For example, we developed multiple codebooks to facilitate analysis of the one-on-one interviews as well as the post-teaching guided-reflection sessions (DeCuir-Gunby, Marshall, and McCulloch 2011). Likewise, we crafted a lesson rubric that was used in our analysis of videotaped mathematics lessons. With the rubric we were able to assess whether teachers' verbal communications during mathematics lessons included elements of SBM and CRP (DeCuir-Gunby, Marshall, and McCulloch 2012). Findings from these various analyses were combined and used to form the teacher case studies that will be discussed later in chapters 4 and 5.

EXTENDING AN INVITATION TO SCHOOLS AND K-2 TEACHERS

Whereas it was necessary to make changes in the NMD research design, our plans for the *participant selection* as well as the *professional development* (project intervention) features of the study underwent what can only be described as radical revisions. Thus, we have written the next four sections of this chapter in a way that offers a fuller account of the original research plan, issues that caused us to discard aspects of that plan, and how these particular parts of our study eventually unfolded. "PLAN A" sections (*italicized*) provide the essence of our *ideas-on-paper*; whereas PLAN B sections detail the actions we undertook (sometimes reluctantly) in the aftermath of multiple modifications and revisions to our original research proposal.

Participant Recruitment: Plan A

Schools (rather than individual classroom teachers) were solicited to participate in NMD. The embedded experimental design for the study necessitated a plan for random selection of schools based on six criteria: (1) Title 1 designation; (2) racially diverse student population made up of at least 30 percent African American students and at least 30 percent White American/ Caucasian students; (3) a minimum of four classes per targeted grade level (kindergarten, first, and second grades); (4) current participation in no more than two countywide initiatives (e.g., textbook adoptions, professional development, etc.); (5) no greater than 50 percent teacher turnover; and (6) school leadership compatibility with NMD focus. By this final criterion we meant school leadership that would support teachers' engagement in instructional experimentation.

In light of strict end-of-year testing protocols that were in place in the school district, it was crucial for the NMD teachers to have the unwavering support of their principals. District leaders helped us to identify a collection of 20 school sites (drawn from over 80 elementary schools within the district) that met the criteria for participation. Our plan was to make school selections in the following manner: five schools designated for the NMD project intervention with Cohort 1 and Cohort 2 teachers; five different schools designated for no project intervention with Control Group teachers; and then five additional schools designated for the project intervention with Cohort 3 teachers. Hence, the goal was to select fifteen schools.

A district official arranged for us to present an overview of NMD at a special meeting with the school principals. The meeting was planned for mid-January but due to a severe snowstorm it was canceled. Consequently, we contacted the principals individually to request a thirty-minute timeslot at their respective schools in February or March to present an overview of NMD. Some principals did not respond to our request; however, the vast majority invited us to attend their next scheduled faculty meeting. In a few cases special meetings were called for the express purpose of our presentation.

We presented the NMD overview to principals and K-2 classroom teachers at 10 different elementary schools. Our goal was to generate interest to such a degree that collections of K-2 teachers (with the support of their building principals) would make a follow-up request to us to be considered for inclusion in the study. Although all K-2 teachers within a participating school would be expected to join NMD, the initial group in the first data collection year (DCY-1) would be limited to just six teachers (two per targeted grade level).

Then, in subsequent years, additional teachers per targeted grade level would be added until by DCY-3, all K-2 teachers within a given school would be "Dreamkeepers." Our goal was to organize three distinct cohorts of participants. Cohort 1, formed in DCY-1, was to be drawn from the original five schools identified through random selection. Across the five schools our aim was to secure a total of 35 participants. For each school this was to include a representative of the school's leadership team (e.g., building principal, assistant principal, or instructional resource teacher) and six classroom teachers (two per targeted grade level).

Each teacher was to be assigned to work with 1–2-grade-level "buddies" during the school year. Also in DCY-1, 30 control group teachers were to be selected to match the Cohort 1 teachers in terms of professional preparation/ degree level, teaching experience, and demographic background. They were to be organized from randomly selected schools drawn from the original 15 that met the qualifying criteria and that expressed interest in joining. Then in DCY-2, Cohort 2 was to be organized. It was to consist of 30 additional classroom teachers drawn from the same five schools used for Cohort 1. The idea was for each school to add two additional teachers per grade level to its corpus of Dreamkeepers.

Finally, at the start of DCY-3, Cohort 3 was to be organized. It was to include up to 60 participants drawn from five additional schools. Among the benefits NMD offered participating teachers were a $1,000 yearly stipend, graduate course credit, district-level licensure renewal points, and (if desired) assistance with preparation for National Board Certification.

Revising Participant Recruitment: Plan B

We closed in on nearly two months of school presentations but had not received a single request from the principals informing us that their teachers wanted to join NMD. At our request, the district's Supervisor for Elementary Programs (DSEP)[3] sent follow-up messages to principals encouraging them to take advantage of the rare professional development opportunity promised by NMD. Surprisingly, her efforts met with modest success.

Only one principal confirmed that teachers from her school might be interested in joining NMD. In a subsequent meeting with the DSEP, we concluded that the brevity of our school presentations might have left some teachers overwhelmed by the various requirements associated with participation in the study. Therefore, together with the DSEP, we came up with a plan to make two more presentations about the project. One would be to school leaders exclusively, and the other would be to classroom teachers.

Our revised presentation to the school leaders occurred in late June and it was attended by a subgroup of elementary school principals participating in an all-day workshop for school administrators that had been previously scheduled by the district. It was our penultimate chance to make a good impression about NMD. Lunch for the attendees was paid for by funds from our grant.

In the revised presentation we incorporated a small group activity in which the principals participated and that was intended to illustrate difficulties many young children encounter with mathematics. Also we invited two of our former graduate students to speak to the principals.[4] With less than two months to go before the start of the school year (and the start of DCY-1), we had been forced to make a major revision in the participant recruitment plan. Gone was the idea of random selection of schools.

Instead, we were now appealing to principals (some of whom had heard an earlier version of our presentation at their own schools) to consider how professional development for their teachers vis-à-vis NMD could work in service to closing the mathematics achievement gap among student groups in their schools. We ended the session by urging the principals to take the idea of NMD back to their teachers and to encourage them to be part of the study.

A draft of a letter we had written detailing what had been covered in the presentation was provided for principals to personalize, copy, and distribute to their teachers. Also, the principals were asked to compile a list (to be retrieved two weeks later by an NMD research assistant, RA) of the names of K-2 teachers from their respective schools who would be attending our second revised presentation.

That presentation, which occurred in August, was incorporated into a "back-to-school" all-day teacher workshop series that had been scheduled several months prior by the district. Over 100 K-2 teachers from throughout the district were in attendance. Box lunches were provided and paid for by the NMD grant. Ultimately, the two revised presentations about NMD yielded a total of twenty-one teachers drawn from three different elementary schools. This group of teachers formed Cohort 1 for the study.

At the close of DCY-1, one project school and over 50 percent of the Cohort 1 teachers were lost to attrition.[5] Even so, consistent with our original plan, additional teachers from the remaining two schools were added to the study. With more recruitment two new schools joined, and our net gain was an additional twenty-four new teachers to comprise Cohort 2. Thus, as we prepared for DCY-2, the participant pool consisted of four schools, four principals (two from DCY-1 and two from DCY-2), and thirty-four teachers.

Then just prior to our official second year kick-off, we learned that the four new teachers from one of our original project schools would not be participating after all. Likewise, approximately six weeks later, one of our newest schools (along with eight more teachers from Cohort 2) left the project.[6] Within two months of DCY-2, Cohort 2 had dwindled to just twelve teachers.

The following year recruitment for Cohort 3 resulted in a bounty of thirty-one new K-2 teachers. This included additional teachers from one of the original project schools as well as teachers from three new schools. Also, we added a fourth new school in DCY-3 whose teachers served as the control group.

It is likely that several factors contributed to the surge of interest in NMD among principals and teachers during that final year. Among these were word of mouth[7] and the fact that by DCY-3 only one year remained in the data collection segment of the project. Over the span of the project's three-year data collection period, a total of ten elementary schools and approximately ninety teachers participated in some aspect of the NMD project.

PROFESSIONAL DEVELOPMENT AS BENEFITS BEYOND RENEWAL CREDITS

The initial plan for the professional development/intervention component of NMD included a two-week summer workshop experience followed by substantive interactions throughout the following academic year between the teachers and school leaders and the project PIs. But not unlike the changes required for the participant recruitment plan, the professional development component of NMD underwent extensive revisions.

Two Intensive Weeks and More: Plan A

The NMD research intervention was to begin for Cohort 1 teachers as a two-week teacher professional development summer workshop (approximately eighty contact hours) prior to the actual start of the school district's academic year. Therein, teachers would be introduced to features of SBM. For example, among the issues to be addressed were differences in instruction that reinforces operations and procedures versus that which promotes deep understanding of broad and transferable mathematics concepts.

Teachers would be challenged to contemplate what it means for a student to "learn" a given mathematics concept. Also, the workshop was to include an introductory exploration of what constitutes culture. A goal was to help teachers think about and begin to analyze critically how elements of culture (e.g., diverse values, worldviews, attitudes) are implicit in every aspect of schooling from classroom behavior policies to assignment grading criteria.

Subsequent to the summer workshop, during the academic year Cohort 1 teachers would be required to observe (and be observed by) one to two Dreamkeeper buddy(ies) teaching mathematics. They were to co-teach lessons weekly, and keep an individual reflective journal wherein they respond to four structured questions formulated by the PI team. Teachers would be required to meet for forty-five minutes weekly to participate in a post-teaching-guided reflection session with an NMD RA.

Finally, teachers were to attend a three-hour weekly cohort professional development session on the university campus (and for which they would receive graduate credit), during which they would engage in professional analysis of dilemmas and problems that had arisen in the context of their mathematics teaching.

For their parts, the designated NMD member of each school's leadership team (e.g., principal, assistant principal, or resource teacher) would be required to attend the two-week summer workshop (along with the school's K-2 teachers); support the teachers' instructional innovation during the academic year; and participate in a monthly two-hour dialogue on K-2 mathematics instruction. In addition to the yearly stipend[8] *(supplemented by a separate stipend for the summer institute), teachers would earn licensure renewal credit through the local school district.*

Project Retreats and Follow-Up Summer Reflections: Plan B

Since we were still trying to recruit schools and organize the first cohort of teacher participants well into the summer, our plan for the two-week intensive workshop was abandoned. The enormity of this decision forced us to reconceptualize the overall plan for delivery of the project intervention. We devised a plan to create a series of six two-day "PD retreats" to be delivered at evenly spaced intervals throughout the academic year.

Each retreat would be held on two consecutive school days (from 08:00 to 16:00), and both days would be structured as a series of interrelated large and small-group learning sessions organized around the study's themes. Cohort teachers would be required to attend both days of each PD retreat. Grant funds would be used to pay to the school district the cost of substitute teachers. Each PD retreat would be held on the university campus and lunch would be provided. The nature of the buddy structure changed. Teachers did not co-teach lessons nor did they maintain reflective journals. They did, however, observe each other teaching mathematics.

Retreat days were organized as one to one and a half hour sessions along with morning, lunch, and afternoon breaks. Sessions consisted of short lectures, large and small group activities, simulations (Shirts 1969, 1977), and work sessions including lesson analyses. Approximately 60 percent of retreat time was devoted to issues and concepts surrounding SBM, whereas 40 percent was reserved for the exploration of issues related to CRP. In some cases more (or less) time was devoted to a particular focus during a particular retreat, and commonly the two foci were merged.

This change fundamentally altered the structure and to some degree the content that had been planned for the PD/research intervention. Instead of it being significantly drawn from questions and issues arising from the dynamic interactions between Cohort teachers and their students, it was revised and largely drawn from the research and anecdotal examples of the PIs' own research and knowledge of the two complementary frameworks (SBM and CRP) that undergirded the study. This reconceptualization of the PD/research intervention, representing a major departure from the original plan for the study, had to be submitted to our grant sponsor for re-evaluation and approval.

Ultimately, the PD retreats were well received and evaluated highly by the Cohort 1 teachers; however, in response to our request for feedback at the conclusion of DCY-1 the unanimous complaint was that there were too many retreats. Moreover, teachers indicated that unless the frequency was reduced they would discontinue participation in the study.

At the start of DCY-2, the number of PD retreats was reduced to four: two in the fall semester and two in the spring semester. These were followed by a two-day "summer institute" session, which consisted of revisiting some of the more complex concepts explored during the span of the academic/data collection year. Due to the two separate cohorts of teachers, retreat sessions had to be differentiated.

Cohort 2 teachers received a series of retreats that were closely aligned with those that had been delivered to Cohort 1 teachers in DCY-1. Alternatively, Cohort 1 teachers received a series of PD retreat sessions that were intended for them to delve more deeply into theoretical underpinnings of SBM and CRP. The PD retreats for Cohort 1 were scheduled on Mondays and Tuesdays, whereas those for Cohort 2 teachers were scheduled for Wednesdays and Thursdays during the same week of those for Cohort 1.

To facilitate interaction between the two cohorts, we organized "cross-cohort intersessions" that occurred from 17:00 to 19:00 on the evening of the first day of the Cohort 2 retreat. During these sessions we organized interactive activities targeting enhancement of the teachers' own mathematics content knowledge or cross-cultural competency. A catered light dinner from a local restaurant was provided as part of each cross-cohort intersession.

In addition to attending a second year of PD retreats, Cohort 1 teachers in their second year as Dreamkeepers were offered the opportunity to "take ownership and promote" what they were learning about the math/culture connection through either making a presentation to their colleagues at their respective schools or participating in an NMD-sponsored mini-conference.

All of the Cohort 1 teachers opted to participate in the mini-conference and in so doing they were required to work jointly with another Cohort 1

Dreamkeeper to present a thirty-minute professional conference-style presentation on how they were incorporating concepts from SBM and/or CRP into their own mathematics teaching.

The date of the NMD mini-conference was advertised in the local school district and it was publicized statewide through the website of the State Department of Public Instruction. Also, information fliers were sent to Colleges of Education in universities around the state. Over 100 elementary teachers drawn from districts from around the state attended the NMD mini-conference.

In DCY-3, a new role was created for Cohort 1 teachers as co-constructors (and co-deliverers) of the project's PD retreats. Each Cohort 1 teacher, in coordination with one to two Cohort 1 peers, was required to deliver a one-hour presentation during each PD retreat session for either the Cohort 2 or Cohort 3 Dreamkeepers. The PD retreats for Cohort 2 were repeats of the more theoretically focused retreats presented during DCY-2 to Cohort 1 teachers; whereas PD retreats for Cohort 3 were repeats of those presented in the first year of participation in the study for Cohort 1 and later for Cohort 2. Table 2.2[9] lists the broad topical foci covered across the four retreats in DCY-3.

Also in DCY-3, a second mini-conference was held in which Cohort 2 teachers were the featured presenters. Like the first mini-conference, it attracted nearly 100 elementary teachers. The cross-cohort intersections were maintained, but in addition a series of online asynchronous conversations was introduced to further facilitate cross-cohort interactions (Marshall 2013).

Table 2.2 NMD Project Retreat Foci

	Culturally Relevant Pedagogy	*Standards-Based Mathematics*
Retreat I	Understanding Culture "BaFá BaFá" Simulation	NCTM Standards Constructing Counting Systems Base_{10} and Base_4 Place Value Systems Counting Strategies
Retreat II	Professionalism: The Dreamkeeper Identity Defining Culturally Relevant Pedagogy	Teaching Mathematics for Understanding Mathematical Models and Tools Learning Trajectories for Addition & Subtraction
Retreat III	High Academic Achievement Language and Cultural Identity	Early Algebraic Thinking Patterning
Retreat IV	Cross-Cultural Competence Ethnoracial Identity Statuses Sociopolitical consciousness	Early Rational Number Sense Fair Shares

AN INTRICATE DATA COLLECTION PROTOCOL

Data collection for NMD took place in multiple venues. For example, teacher mathematics surveys were completed during the first PD retreat session; whereas kindergarten interviews took place right outside of classrooms in school hallways. The most expansive aspect of data collection, however, was in videotaping mathematics lessons and facilitating and recording post-teaching-guided reflection sessions. These two data sources were time intensive and required tremendous coordination of multiple schedules.

Project RAs were required to collect four data sets for each teacher. A data set consisted of two consecutive lessons (per teacher within a buddy pair/triad) plus one post-teaching-guided reflection session. All members of a same-grade buddy pair/triad attended the same reflection sessions, which were structured around four primary questions. The four prompts used in each post-teaching reflection session were: (i) What specific mathematical ideas were the students supposed to learn; and what does it mean for someone to understand these ideas? (ii) What understanding DID the students actually demonstrate? How do you know? (iii) What were you doing that promoted or hindered the learning sought? (iv) What would you teach/do in the next lesson(s)? Why?

The questions prompted teachers to think critically about and articulate their instructional intentions vis-à-vis the mathematical ideas incorporated into their lessons. Additionally, teachers were prompted to discuss the outcomes of their lessons in terms of actual student learning. Because buddy teachers observed each other's lessons, they were expected to share critical reflections on their own as well as each other's lessons. The average duration of a lesson was thirty to forty-five minutes, whereas guided reflection sessions among NMD buddies facilitated by an RA lasted on average from one to one and a half hours. One-on-one interviews with each teacher were conducted at the start of each data collection year and then again at the close of the year.

By the conclusion of the study, the corpus of data included approximately 500 videotaped mathematics lessons, 118 videotaped post-teaching reflection session discussions, 134 audiotaped teacher interviews, 16 audiotaped principal interviews, 223 kindergartner mathematics assessments, 222 first-grader mathematics assessments, 195 second-grader mathematics assessments, and a host of retreat session artifacts.

THE NMD RESEARCH TEAM

Completion of NMD required tremendous coordination and flexibility along with the commitment of many individuals. Our research team was a tight-knit crew of varying, albeit largely complementary, personalities and temperaments. Each individual brought a reserve of professional knowledge, skill, and talent that was critical to the project's completion. From its earliest inception to its conclusion the team demonstrated commitment to diversity.

Among the dimensions of diversity apparent in the team's composition were ethnoracial, gender, and scholarly focus. The extended research team started out with twelve members including the project PI, co-PI,[10] and five graduate RAs (two doctoral students and three master's students; three Black, two White; two females; three males).

Three of the RAs (one female/two males) were mathematics education majors whereas the other two (one female/one male) were elementary education majors. Also, there was a five-member national advisory board (three females/two males) that included experts in mathematics education research and multicultural education research. Each member of the NMD team was deeply committed to the success of the project and they worked individually and in tandem toward realizing its goals.

The first year of the project was devoted to preparing the RAs. To this end the initial meetings laid the groundwork for building *esprit de corps* among the members of the team core (PI, co-PI, and RAs). Once this was in place, RAs were required to complete collections of scholarly readings related to the CRP and SBM foci of the project, including empirical studies as well as theoretical and conceptual articles and chapters written by leading scholars in both fields. Weekly three-hour sessions were held in which the team core members engaged in critical discussion of the readings along with explorations of connections to and implications for key features of the project.[11]

By the end of the first semester, the seven-member research team core had coalesced successfully around working toward successful completion of the NMD project. An apt metaphor to describe the interactions among the members of the core was research team as loving family. However, at the start of the second semester of the project (weeks prior to the scheduled January meeting with the principals), a tragic event occurred that pierced the team at its core.

For the next twelve months reverberations of that event weighed heavily on all of the members, and by the close of DCY-1 the original research team core collapsed and was eventually disbanded.[12] Subsequently, the core was rebuilt[13] and the project proceeded on its course. Over the span of the study, some thirty-one individuals served as members of the NMD research team. Among the team roles were co-PI for mathematics, co-PI for multicultural studies, co-PI for research methods, subject matter consultant, advisory board member, graduate RA, data collector, and webmaster.

AN AUTHENTIC STUDY

The NMD project was an enormous undertaking. This chapter described how the NCLB legislation was a critical impetus for the study. We detailed aspects of the original research agenda, and disclosed issues that necessitated its modification and revision. In the process of organizing this chapter, we made the bold decision to share a few of the more noteworthy missteps that shaped the course of NMD. Correspondingly, we outlined recovery efforts we undertook as researchers to keep the project on course. Without these details, the story of NMD would be wholly incomplete.

Ultimately, our goal in this chapter has been to present a true-to-life picture of the complexities of conducting research in a *real-world* context. In our case the context was represented by the dynamic and often unpredictable world of public schools. The NMD project was an authentic research study presented with many external challenges. Its testament, however, is how it survived its own internal trials. In the remaining chapters we delve into the outcomes of the study, and we examine its implications for future research.

NOTES

1. We recognize that debates about the reauthorization of NCLB are currently underway. Our use of past tense is simply in keeping with explaining one impetus for formulating the original proposal for the NMD project.
2. Some perceived that the sanctions imposed on schools were excessively punitive. For example, failure to meet yearly progress goals for even one targeted population group could result in a lowered ranking for an entire school that formerly had been designated as high performing. Another issue that resonated among many was the paucity of resources available to districts populated with high percentages of children from the targeted populations.
3. Position titles have been altered to maintain the anonymity of the individuals who participated in NMD.
4. One of the former students was working as an assistant principal and the other held an Instructional Resource teacher post. Each offered a "testimonial" about the quality of work we as co-PIs of NMD had done previously in our respective fields (i.e., multicultural education and mathematics education). They spoke of graduate courses as well as teacher workshops each had conducted.
5. All but one of the Cohort I teachers (i.e., twenty out of twenty-one) expressed strong interest in returning for a second year in NMD; however, in addition to the one teacher who left for "personal reasons," ten others had to leave the project at the close of DCY-1. Among the other reasons were: involuntary transfers to other schools within the district, relocation to new districts in the state, relocation to neighboring states, and one advanced stage pregnancy. In addition, just prior to the start of DCY-2, one Cohort 1 school received a new principal who subsequently decided not to allow the two Cohort 1 teachers from that school to continue participating in the study. He explained that because a group of first year teachers had been hired in the school's primary grades, the more experienced teachers (Cohort 1) from that school would be needed to serve as mentors.

6. Their departure was due in part to conflicts with the NMD data collection plan. The principal at the school, with the support of the teachers, had implemented a schedule in which mathematics lessons were taught at the identical time in all K-2 classrooms.

7. We recruited participating teachers from both Cohorts 1 and 2 to share testimonials of the benefits they had gained from participating in NMD during the third year of recruitment.

8. Stipends were determined by the NMD funding agency guidelines. Teachers received $1,000 (paid in two installments) during the nine-month academic year, and $250/day for participation in summer activities.

9. Different individuals served as co-PI for mathematics during the three-year data collection period. Consequently, there were slight changes in the focus of mathematics activities between DCY-1 and DCY-2, as well as between DCY-2 and DCY-3. The CRP focus was slightly different from DCY-1 to DCY-2 owing to the decrease in the number of retreats from six to four, but otherwise its focus for the three-year period was consistent.

10. Due to the various demands of the project vis-à-vis the PD delivery, just prior to the start of DCY-1, the two-member PI/co-PI team discussed adding a consultant for research methodology to the research team. This change eventually did occur at the start of DCY-2. See endnote 13.

11. RAs earned three hours of graduate credit as members of the NMD research team. In addition, the grant covered university tuitions, and each RA was paid a monthly salary.

12. The tragic event was the untimely death of the daughter of the project's first PI. It had a devastating effect on the entire research team. By the start of the second semester of DCY-1, the original PI made the decision to sever all ties with the project prompting one RA to say, "I feel like my parents are divorcing." Cohesion among the RA team began to falter, and at the close of DCY-1, four of the original RAs were dismissed for varying levels of insubordination.

13. During the span of the project, the composition of the senior research team (i.e., PI/co-PI) underwent several revisions. Immediately after the resignation of the original PI, the original co-PI and lead author Pat Marshall assumed the role of project director and lead PI. Subsequently, two new co-PIs and two consultants were added to the team. Also, a new group of graduate student RAs was hired. As the project progressed, eventually two retired teachers and two undergraduate students were brought on board to assist with data collection and initial-level data analysis. As authors of this book, we were the third (and final) NMD PI/co-PI team. Pat Marshall, whose area of research is multicultural studies, was coauthor of the original NMD research project proposal and she was the original co-PI for the study. At the start of DCY-2, a second co-PI (for mathematics) joined the research team, along with Jessica DeCuir-Gunby, who was added as co-PI for research methodology (see endnote 10). Then, the summer before the start of DCY-3, Allison McCulloch joined as the third co-PI for mathematics.

Chapter 3

Math Learners and Cultural Beings

K-2 Teachers as Dreamkeepers

The participant pool for NMD consisted of eleven elementary school leaders and sixty-five primary grade teachers.[1] All were female except for one male school principal. Each contributed to the overall success of the project; however, we have chosen a multiple case study approach (Yin 2004) to detail the findings and implications of the project as well as to offer a fuller portrait of the distinctiveness of a select group of the participants. Specifically, in this chapter we introduce the subset of nine NMD project teachers whose mathematics instruction vis-à-vis CRP will be detailed and analyzed throughout the remainder of the book.

We begin with an overview of the broad context in which the teachers worked including the sentiment surrounding equal access to high-quality pre-collegiate education in the state, characteristics of the local school district, and features of the individual project schools. Next we describe the leadership within each project school and provide a brief account of each school leader's involvement with NMD. Then, we discuss the three teacher cohorts by highlighting the "personality" associated with each. Finally, we present a brief profile for each of the nine case study teachers.

THE CRUTCHFORD-WHINTON SCHOOL DISTRICT

Several years before the start of NMD, a prominent state superior court judge issued an order that reverberated around the region. The judge had reprimanded officials in four low-wealth school districts where academic underachievement and school dropout rates were among the highest in the state. Responding to a lawsuit from parents of students in the four districts, the judge ordered that significant changes be made to improve academic achievement.

At the core of the ruling was a previously obscure clause in the state's constitution, which guarantees every child "the right to an equal opportunity to obtain a sound basic education." Echoes from the judge's ruling inspired us to invite one of the larger districts in the state, Crutchford-Whinton, to participate in NMD. It had not been among the districts cited in the judge's court order, but the academic achievement gap between Black and White students in the district was very well known.

The Crutchford-Whinton School District (CWSD) covers a vast area that includes a medium-sized city along with several other smaller jurisdictions. Since the 1990s, a steady stream of newcomers—including "transplants" from all over the United States—have been relocating to the area. Likewise, Crutchford-Whinton attracts immigrants from countries throughout the world and whose statuses range from undocumented to highly credentialed.

The children of this diverse assortment of transplants and immigrants attend elementary and secondary schools in CWSD.[2] Possibly unknown to many of the newcomers is the fact that the geographic expansiveness of CWSD is owing to the mid-1970s merger between the then (mostly Black) Crutchford school district and the then (mostly White) Whinton school district.

Business and political interests converged around the issue of public schools to usher out the infamous "separate-but-equal" policy that had long been in effect in the area. The primary reason for the convergence was to curb "White flight," which was on the rise in reaction to desegregation plans throughout the region. Since that time, the CWSD has enjoyed a string of accolades that are the envy of other districts in the state, if not the country. The student demographic profile of CWSD has changed considerably in the last twenty-five years and correspondingly, the "achievement gap" has become more complex. Yet it is often treated as little more than a pimple on an otherwise baby smooth complexion.

THE PROJECT SCHOOLS

Recruitment of teachers to participate in NMD began at the school level. Indeed, as detailed in chapter 2, it was schools, rather than individual teachers, that were the target of our recruitment efforts. A total of ten schools participated in NMD during the span of the project, therefore we use the term "project schools" in reference to those that were still part of the study at its conclusion.

By the end of NMD, there were six project schools and one comparison school. The teachers chosen for the case studies are drawn from four different project schools—Rhine River, Farren Reese, Dawson Dome, and Juniper. Descriptions of the schools are based on PI and RA field notes as well as teacher and principal interviews.

Rhine River Elementary

Rhine River is located on the outskirts of the county in a suburban area. The school is situated in the middle of a residential community of single-family homes that are modestly priced for the Crutchford-Whinton area. It has a culturally diverse student population comprised primarily of Whites (66.9 percent), African Americans (15.8 percent), and Hispanics (10.3 percent). Additionally, at the time of the study 25 percent of the students at Rhine River were eligible for free or reduced-price lunch while approximately 8 percent did not speak English as a first language.

During the project, Rhine River was converted from the traditional nine-month school year calendar to a year-round schedule.[3] Because of this change, many parents living in the neighborhood that surrounds Rhine River opted to enroll their children in other CWSD schools in order to remain on the traditional nine-month calendar.

A banner is displayed prominently at the front of the school that indicates Rhine River was once an "Honor School of Excellence." The school itself feels very inviting with a carpet that says "Welcome to Rhine River Elementary" placed at the main office doorway. Along a hallway is a showcase featuring the school's distinguished teachers. Various pictures are posted and they include captions indicating the activities and leadership roles each teacher assumes in the school community. The NMD teachers are well represented with some appearing in multiple categories. All of the hallways feature student work including art, language arts, and mathematics.

Like most elementary schools, Rhine River is organized by grade levels. Kindergarten and first grade classes are located on one hallway while the second and third grade classes are grouped together on another. At the start of NMD, most of the fourth and fifth grade classes were located in trailers located adjacent to the main school building. This is because prior to its conversion to the year-round calendar, space was at a premium in Rhine River. Although several classes were taught in trailers, more were needed to accommodate the increase in the number of children attending the school.

During its second year in the project, the Rhine River population increased by 100 students. As a result, teachers and the school leadership displayed a high degree of creativity in the use of space. For instance, a few teachers taught classes in the cafeteria or the media center. A closet was converted to a workspace for small group instruction. The need for space decreased after the year-round calendar conversion.

Rhine River was the most unique of the NMD schools in that by the close of the project it was the only one that still included teachers from all three cohorts. A total of eleven of the project teachers were from Rhine River, including two of the primary case teachers (Johnetta and Giselle) and one of the secondary case teachers (Aisha).

Farren Reese Elementary

Farren Reese is located in a suburban area on the outskirts of the eastern part of the CWSD. At the time of the study it was one of the fastest growing areas of the county as was evident by the fact that there were two other elementary schools located in close proximity including one less than a mile away.

The Farren Reese building was just three years old when the school joined the study. Its paint was fresh and the environment was clean to the point that it felt sterile—almost like a doctor's office. Even so, the building looked and sounded like a traditional elementary school with colorful paint, pictures on the walls, and the chatter of high-pitched voices. The student population was ethnoracially diverse with a near equal percentage of African Americans, Latinas/os, and Whites.

Walking through the hallways, samples of children's work are on full display. In addition, commercially produced posters featuring words associated with character traits are visible throughout. They remind the children to "Respect," "Care," "Always Be Safe," and "Excel in All You Do." Individual classrooms display character trait posters as well. Some unique posters hanging on the walls along the hallways include information about saving the planet and maintaining a healthy heart. Some posters are in Spanish.

There seems to be a variety of resources. For example, the art room has an abundance of supplies and the music room has a piano as well as other instruments. The media center features a variety of books and new Dell computers. A large gymnasium has six basketball hoops. Then there is the cafetorium (combination of a cafeteria and an auditorium) with a school supply store where students can purchase items at an inexpensive price. A spacious lounge with comfortable chairs (as well as snack and drink machines) calls out to teachers at break times.

Farren Reese participated in NMD all three years, although by the end of the study, the remaining six teachers from that school were all members of Cohort 1. There had been four teachers from this school who joined Cohort 2 but their participation ended very early in DCY-2. Similarly, in DCY-3 one Farren Reese teacher joined Cohort 3, but she too subsequently left the study for personal reasons. One of the primary case teachers (Arlene) and two of the secondary case teachers (Kendra and Tasha) were from Farren Reese.

Dawson Dome Elementary

Dawson Dome is located in the southwestern part of greater Crutchford-Whinton, another area that at the time of the study was experiencing rapid population growth. It is situated on a large campus that shares space with a middle school. The surrounding residential area includes both older and newer single-family homes as well as a fairly large apartment complex. Within a mile of the school is a large shopping mall.

Upon entering the school, the left side of the brightly lit hallway features windows that line the main office, whereas the right side has a beautifully painted mural of an outdoor nature scene. There is a large quilt hanging above the office windows. It looks to be made of student drawings representing different regions of the state.

Just below the quilt is a framed U.S. flag that was dedicated to the school. The floor is painted with large blue and yellow stripes and stars. For its part, the main office looks very inviting with potted plants outside the door and several yellow chairs and a loveseat in the waiting area. The waiting area is usually decorated for the seasonal holiday (e.g., turkeys for Thanksgiving, a decorated tree for Christmas, etc.).

Dawson Dome is divided into wings with one dedicated to the kindergarten and first grade classrooms, and the other to the second and third grade classrooms. Fourth and fifth grade classrooms are located in the center of the school along with the gym, media center, and cafeteria. Also there are trailers located on the periphery of the building. The walls of the various hallways showcase the work of the children.

The school's population is representative of the surrounding area in that it is ethnoracially diverse. During the span of the project, numerous students enrolled in and transferred from Dawson Dome, giving a sense that the school's overall population was quite transient. Many of the students were learning English as a second language (ESL). At the same time, the number of ESL teachers at the school was clearly inadequate to serve the size of that part of its population.

It seemed possible that the paucity of ESL teachers may well have been an inadvertent contributing factor to the student transience. Dawson Dome joined the study in its final year with five teachers participating. One (Isabel) is included as a primary case teacher.

Juniper Elementary

Juniper is located in an older residential neighborhood near the downtown area of the medium-sized city. The school building was recently renovated and as a result the windows and handrails are noticeably newer than the structure itself. At the front of the building is the sign "IB World School." It is shorthand that lets everyone know Juniper is an International Baccalaureate Spanish Language Magnet school.

Many of the decorations at the front entrance reflect this distinction including the flags from different countries, which hang from the ceiling near the main office. There are also commercially produced posters written in different languages displayed on the walls.

Near the main entrance is a showcase of students' awards and academic work, including a "Writers' Wall of Fame" that highlights the students' compositions. The main office has big windows that look out to the school's entryway allowing arriving visitors to see inside. It is usually a busy place with myriad activities such as teachers making copies, students getting bandages, and parents waiting to pick up children. A computer lab, the library, and administrative offices all branch off from the main entryway.

As visitors make a right turn out of the main office and proceed down the short set of stairs, they can encounter either of two hallways. The left one leads to the primary grades. Outside the kindergarten, first, and second grade classrooms student work is displayed, almost like wallpaper. The right hallway leads to the area for "upperclassmen." Curiously, substantially less student work is on display outside the third, fourth, and fifth grade classrooms.

Even so, the upper grade classrooms themselves are filled with symbols of inquiry. Art and music rooms are located on this hallway and they feature vibrant mosaics and pictures. The school has a separate building where the cafeteria and a gymnatorium (combination gymnasium and auditorium) are located. Also, there is a large playground with two sets of play stations and a soccer field. Like Dawson Dome, Juniper joined the project in DCY-3. There were twelve teachers from Juniper all in Cohort 3. Included among them are two of the secondary case teachers (Hadley and Jenna).

LEADERSHIP IN THE PROJECT SCHOOLS

The leadership team within each project school was critical to NMD. All of the school leaders were committed to high-quality mathematics teaching. Similarly, most expressed appreciation for the effects of culture on the teaching-learning process broadly, if not on mathematics learning specifically. School leadership teams were expected to be demonstrably supportive of the K-2 teachers' participation in NMD. Among other things this meant encouraging teachers to try out features of the project's professional development intervention in mathematics instruction.

In addition, at least one member of a school's leadership team (e.g., principal, assistant principal, instructional resource teacher) was expected to participate in some of the NMD project activities. (See chapter 2 for a discussion of the original plan for school leadership involvement in NMD.) Hence, to a significant degree, principals (and/or their designates from the school's leadership team) made important contributions to the NMD experiences of the cohort teachers at their respective schools. The next sections provide a brief overview of the leadership team designate from each project school.

Ms. Felicia Williams, Principal of Rhine River

The principal for Rhine River, Ms. Felicia Williams, can be described as both active in NMD and consistently supportive of the teachers from the school who participated in the project. Actually, Rhine River joined the project under the direction of a different principal. When the first principal transferred from Rhine River to lead a brand new school in the CWSD, Ms. Williams became principal of Rhine River.

Ms. Williams immediately showed support for Cohort 1 teachers' continuing their participation in NMD under her leadership of the school. Likewise, she encouraged a second group (Cohort 2) to join the project. Ms. Williams attended the first retreat of DCY-2 and she participated intermittently in other project events including the mini-conferences (see chapter 2). Therefore, in a subsequent one-on-one interview, it came as a surprise to learn that initially she had felt reluctant to sustain Rhine River's participation in NMD.

The project PIs had not detected the slightest hint of reluctance from Ms. Williams about Rhine River's continued participation in the project. Nor, apparently, did any of the project teachers from the school. Indeed, one teacher described both Ms. Williams and her predecessor as "math persons" who encouraged teachers to participate in NMD. By DCY-3, Ms. Williams even actively promoted extension of the project by allowing project teachers at Rhine River to plan and implement a professional development afternoon workshop session (based on ideas drawn from NMD) for their non-NMD colleagues at the school.

Ms. Dolores Perkins, Principal of Farren Reese

Dolores Perkins was a complicated leader of Farren Reese in that teachers at the school sensed she gave mixed messages about their involvement in NMD. At the same time, Ms. Perkins seemed to genuinely appreciate the goals of NMD and being partnered with the university through the research project.

Ms. Perkins attended some of the project-sponsored events (e.g., mini-conferences, Next Steps Meeting) and was an active participant. By DCY-3, however, teachers from Farren Reese reported inconsistency in the level of support they received from Ms. Perkins. For example, some Cohort 1 teachers expressed exasperation over being assigned additional responsibilities at the school during DCY-3 despite the fact that by then they were required to co-plan retreat activities with the project PIs. Others complained that Ms. Perkins required them to be involved in several new initiatives in addition to their involvement in NMD.

Every teacher from Farren Reese shared that Ms. Perkins often extolled the virtues of their participation in NMD to school district officials. Thus, based on reports from some teachers, in her enthusiasm for high-quality professional development opportunities for teachers at Farren Reese, Ms. Perkins may have unknowingly contributed to a sense of overwhelm among some teachers. In this regard, by DCY-3 at least some Farren Reese teachers perceived Ms. Perkins to be sending mixed messages regarding her support of their full participation in the project.

Ms. Mary Lewis and Ms. Tracey Rhodes—Principal and Assistant Principal of Dawson Dome

Mary Lewis, principal of Dawson Dome, was among the most enthusiastic school leaders at our presentation to recruit Cohort 3 schools and teachers. Moreover, she shared that although she had been aware of (and had heard good things about) NMD in previous years, it had not been until DCY-3 that she was prepared to commit the school to involvement.

Despite Ms. Lewis's enthusiasm for NMD, it was the assistant principal, Tracey Rhodes, who assumed the most active role as the school's leadership liaison to the project. This was a logical choice since Ms. Rhodes was the member of the Dawson Dome leadership team who was in charge of the K-2 grade levels at the school. It was she who typically served as the official representative from Dawson Dome at NMD events.

Both principal and assistant principal desired to implement long-term changes at the school vis-à-vis mathematics teaching, and they perceived the goals of NMD to be compatible with their own views of learning. Teachers at the school joined NMD as a direct result of a collective concern of the school's leadership team over improving students' mathematics learning.

In addition, Ms. Lewis and Ms. Rhodes perceived NMD would provide the necessary support to help close the achievement gap between student groups at the school. Hence, as a leadership team they encouraged the participating teachers of Dawson Dome to share knowledge learned from the NMD project with other teachers in their school.

Mr. Charles Keegan—Principal of Juniper

Charles Keegan recommended teachers at Juniper participate in NMD because he was keenly interested in encouraging them to learn new ways to teach mathematics. He was a big proponent of collaboration and reflective practice and believed that in order for the teachers to become better teachers of mathematics, they needed to learn strategies from each other as well as take time to reflect upon their own teaching practices.

In addition, Mr. Keegan felt that in order to develop into better mathematics teachers, they needed to learn multiple ways to teach mathematics concepts and determine how to differentiate mathematics instruction for diverse learners. As such he sought to provide teachers at the school an innovative professional development.

Showing no reluctance to take advantage of the opportunity offered by NMD, he strongly encouraged as many Juniper K-2 teachers as possible to participate in the study. In fact, with a total of twelve teachers, Juniper had the largest number of participants to join the project in any single data collection year.

Mr. Keegan especially liked the NMD emphasis on collaboration, reflective teaching, and the sharing of best practices. Through his support of NMD, Mr. Keegan continuously provided encouragement to the Juniper teachers and readily allotted the necessary time for them to participate in project activities. Like the other principals, he attended a variety of NMD events and thought positively of the project.

THREE TEACHER COHORTS

Recruitment for NMD occurred at the school level with the process itself taking place over three consecutive academic years. In conjunction with the original quasi-experimental design of the project, we organized the teacher participants into cohorts to examine the impact of the project retreat intervention. Our goal was to assess the impact in light of the duration of the teachers' participation in NMD. Therefore, the cohort structure was a key component of the NMD design.

Teachers of each cohort not only began their tenure in the project the same year, but they were encouraged to establish multidimensional bonds with each other based on the following features: same grade/same school; cross grade/same school; same grade/cross school; and cross grade/cross school.[4]

The intricacy of the bonds inevitably promoted a degree of closeness such that by the conclusion of the project, three distinct cohort personalities had emerged. The following sections provide a general overview of the compositions of the three cohorts at the close of the study.

Cohort 1: The Leaders

The teachers of Cohort 1 joined the project in September 2005. They were eight experienced primary level teachers, (i.e., two kindergartens, four first grades, and two second grades) from two project schools (Rhine River and Farren Reese). Their group was racially balanced between Blacks (50 percent) and Whites (50 percent).

When they joined NMD, the average teaching experience for Cohort 1 was approximately thirteen years, with the average tenure at the schools being three years. For most of the project the highest level of formal education completed by Cohort 1 teachers was the bachelor's degree, although one member held a master's degree.[5] Seven of the Cohort 1 teachers are originally from the South; one is a Northeastern transplant having come to the South several years prior to NMD.

In general, Cohort 1 members spring from working-class backgrounds. When describing their own mathematics histories, nearly all reported that their experience in learning mathematics changed for the worse in either middle (junior high) or high school. Similarly, all reported consciously taking the fewest mathematics courses necessary to complete degree requirements. Six Cohort 1 teachers taught at Farren Reese and two taught at Rhine River.

Overall, Cohort 1 teachers saw themselves, and were seen by other teachers in the study, as the leaders in the project. We believe this "cohort personality" was facilitated by the feature of the project intervention that required them to assume a different, albeit more complex, professional role (Sprinthall and Thies-Sprinthall 1983)—giving back to the professional community—after their first year of participation.

In DCY-2 they were required to make presentations as part of the mini-conference, whereas in DCY-3 they were required to joint-plan retreat activities with the project PIs. Similarly, several Cohort 1 teachers took on leadership within their respective schools. As a whole, Cohort 1 teachers saw themselves as "Dreamkeepers."

Cohort 2: The Friends

The members of Cohort 2 joined the project at the start of the 2006–2007 school year. They were ten teachers drawn from two project schools (Rhine River and Wolfson): seven from the former and three from the latter. Like Cohort 1, the group was ethnoracially diverse in that it consisted of four Blacks, five Whites, and one Asian. Most had southern roots ($n = 7$), although two participants were reared in the Northeast and one was raised outside the United States.

On average, they had over eleven years of teaching experience and had taught at their respective schools for an average of six years. Six held bachelor degrees while four held master's degrees. Teacher grade level distribution among Cohort 2 was relatively even with four teachers at the kindergarten level, three at first grade, and three at second grade.

Cohort 2 teachers came to the project with a wide range of mathematics backgrounds. Several of the teachers described themselves as comfortable with mathematics while many others shared that they did not like, or were apprehensive about, mathematics because of past experiences.

Overall, Cohort 2 can be described as the liveliest and most energetic of the three cohorts. This may be attributed to the fact that several of the teachers at Rhine River were friends and socialized with each other outside of the work context. We believe that as a consequence of existing friendships, Cohort 2 teachers were more receptive to challenging each other's thinking and personal stances in relation to some of the more challenging issues explored in conjunction with CRP, including racial identity development.

Similarly, the individual personalities of Cohort 2 teachers were quite different from those in Cohort 1. For example, in the former group there were several outspoken members, whereas in the latter (Cohort 1), the individual personalities were more reserved. Likewise, although Cohort 1 became evenly divided racially, it began as a largely White group. Conversely, the racial balance of Cohort 2 from the outset was more balanced. Cohort 2 retreats were often filled with more talking and laughter with the individual members of the group engaged in playful banter.

Cohort 3: The Onlookers

As the last group of teachers to join NMD the members of Cohort 3 were different in several ways from the previous two cohorts. They joined at the start of the 2007–2008 school year, and from the outset they were acutely aware that their experience would be different from the teachers of Cohort 1 and Cohort 2. Interactions between Cohort 3 teachers and the research team members seemed less easy and not as free flowing as interactions with the other cohorts. The brevity of their tenure in the project made it more challenging to gain substantive insights into the personalities of individual teachers.

Cohort 3 consisted of thirty-one women from six different schools (Farren Reese, Rhine River, Wolfson, Dawson Dome, Juniper, and Woods). Unlike Cohorts 1 and 2, the distribution of the teachers among the six schools that comprised Cohort 3 was starkly uneven. There was one teacher from Farren Reese, two from Rhine River, two from Wolfson, five from Dawson Dome, twelve from Juniper, and nine from Woods.

The evenly distributed ethnoracial diversity of Cohorts 1 and 2 was absent in Cohort 3. Out of a total of thirty-one members, two were Black and twenty-nine were White. Further, Cohort 3 teachers had the least amount of teaching experience averaging nearly eight years with just over three-and-half years at their respective schools. Grade level distributions among Cohort 3 reveal that twelve members taught kindergarten, seven taught first grade, and twelve taught second grade.

The personality of Cohort 3 as a collective can be described as observant yet cautious. Unlike Cohorts 1 and 2, members of the final cohort had far less opportunity to grapple with the many difficult topics that were introduced in the project. Likewise there was far less cultural diversity and experiential variation among the members of the group. Indeed, an undercurrent of resistance to examine critically the relationship between culture and the teaching-learning process, or what Irvine (1991) characterized as cultural aversion, was apparent among members of the group.

Some Cohort 3 teachers, for instance, were discomforted by discussion of race in the project retreats preferring a color-blind stance in relation to the growing diversity and the achievement gap in the CWSD schools. Others were impatient with explorations of issues related to standards based mathematics, preferring "make-and-take" activities to those in which they were required to assess their own knowledge and critical thinking. Despite this, when asked to reflect on their experience in NMD, all Cohort 3 teachers indicated that the project had helped them to become better mathematics teachers.

INTRODUCING THE MATHEMATICS DREAMKEEPERS

Inevitably in a study that extends multiple years, some of the participants will stand out from the rest as exemplars, positive or negative, vis-à-vis aspects of the study's objectives and findings. Often, stand-out research participants exhibit unique qualities, and they may have particularly noteworthy personalities. Similarly, they may provide especially poignant examples of particular issues that arose in relation to the study. And so it was with the participants in NMD.

We identified nine teachers to serve as case studies that will be used to illustrate the findings from the project. Four of the case study teachers have been designated as primary cases and the other five as secondary cases. The case study teachers were drawn from all three cohorts and all three of the project grade levels are represented within the group. Each case study teacher taught at one of the project schools profiled earlier in this chapter. All Cohort 1 cases participated in the study for three consecutive years, Cohort 2 cases for two consecutive years, and Cohort 3 for one (the final) year. Four of the cases are from Cohort 1, two from Cohort 2, and three from Cohort 3.

Selection of these nine teachers was based on careful analysis of multiple data sources, including cultural and mathematics autobiographies, beginning and ending scores on the Teacher Mathematics Questionnaire (TMQ), and one-on-one interviews. Also, selections were based on evidence of culturally relevant teaching practices as determined by scores on the NMD lesson analysis rubric.[6] Our goal was to identify cases of teachers whose pedagogy during the span of the project was most consistent with the main ideas promoted in the professional development retreats, and most representative of the larger themes reflected in our analyses.

Ultimately, the pedagogy of the four primary case teachers supported the overarching theoretical framework for the study, whereas the five secondary cases uniquely challenged some of our most fundamental theoretical conceptions. In the remaining sections we provide brief introductions to each of the nine case study teachers.

The Primary Cases

In alphabetical order our primary case teachers are Arlene, Giselle, Isabel, and Johnetta. As a collective this group was distinguished for both the manner in which they engaged in the project retreats as well as how they drew upon the two project foci to enhance their pedagogy. Below are brief overviews of each.

Arlene

Arlene was an experienced second grade teacher at Farren Reese. When she joined NMD, Arlene had been teaching for twelve years. She is White and was a member of Cohort 1. Arlene was one of the livelier members of her cohort in its first year. She had a lot to say about many subjects and easily participated in activities. Nevertheless, occasionally she would publicly announce that she was self-monitoring.

During a casual conversation over lunch at one of the first year retreats she shared that as an elementary student she was called "motor mouth." Arlene joined the NMD project based on a recommendation from her principal and she was interested in pursuing National Board Certification. She was born and raised in the South, and describes herself as having been reared in a family that started poor but became middle class.

Giselle

Giselle was a White, Cohort 2, second grade teacher at Rhine River. At the time of the study, she had been teaching at Rhine River and professionally for six years. She initially became interested in the project because of a recommendation by her principal and because she was interested in pursuing National Board Certification. Giselle felt very comfortable with the mathematics aspect of the project in that she was a self-proclaimed lover of mathematics. However, she experienced some difficulty with the culture component.

Giselle grew up in a racially and socioeconomically homogeneous environment. Her small, middle to upper middle class town was majority White. She was not exposed to ethnoracial diversity until college; even then, her exposure was limited. She had difficulty thinking about herself in terms of culture and did not view herself as having a cultural identity. She even struggled with interacting with other teachers at her school, as well as other teachers in the project, because of cultural differences.

Isabel

Isabel was another experienced teacher that taught elementary school for twelve years. She was a White, Cohort 3, first grade teacher at Dawson Dome. She joined the NMD project because it was highly recommended by her school principal. In addition, she was interested in becoming National Board certified and thought that participating in a project with an emphasis on mathematics would be helpful in allowing her to accomplish this goal. Although her reasons for joining the project centered on mathematics, Isabel did have an interest in culture because of her childhood experiences.

She grew up in a northern state in a predominantly White community, attending mostly White schools (both students and teachers). Her parents divorced when she was five years old, causing her mother to struggle financially. Because of this, Isabel was always sensitive to issues regarding socioeconomic status (SES) and easily related to her low-SES students. The NMD project helped her to further her interest in culture, particularly learning more about the relationship between race and SES.

Johnetta

Johnetta taught first grade at Rhine River; she is African American and displayed a positive and largely confident professional demeanor. The latter is partly evidenced by the information that appeared in her electronic signature block. It informed all recipients of her emails that she is a National Board Certified teacher.

Throughout her three years in NMD Johnetta maintained a focus on the study's questions surrounding the impact of culture on the teaching-learning process although near the end of the first year she had begun to exhibit acute interest in mathematics as well.

Her interest in the NMD project grew stronger and she became more committed to its dual focus. Earlier in her professional career Johnetta taught in a small rural school district. That area was distinguished by the fact that as late as the 1980s Black/White race relations had not changed significantly from the 1960s when Johnetta had been an elementary student.

The Secondary Cases

Although many common themes were found among the experiences of the participants as demonstrated by the primary cases, our remaining five case study teachers had unique stories to tell. From their stories we extrapolate distinct themes that offer important insights about the study overall. The five teachers who were selected to be the secondary cases are Aisha, Hadley, Jenna, Kendra, and Tasha. These teachers varied widely in relation to their interpretations and reactions to the experience of NMD. Below are brief profiles of each.

Aisha

Aisha was an experienced, African American, Cohort 2 kindergarten teacher from Rhine River. She joined the NMD project because of her previous acquaintance with one of the project PIs as a former instructor. Even so, she was excited about the opportunity to learn more about the topic of culture and its impact on the teaching and learning of early mathematics.

Because of Aisha's knowledge of multicultural issues, she was very comfortable with cultural diversity conversations. She was not afraid to ask critical questions of her colleagues, particularly her fellow Cohort 2 members from Rhine River. The NMD project was the perfect opportunity for her to share her considerable knowledge of multicultural education and to enhance her mathematics teaching skills.

Hadley

Hadley was an experienced second grade teacher at Juniper. When she joined NMD she had been teaching for thirteen years. She is White and was a member of Cohort 3. Hadley was a vocal member of this, the largest, cohort. Hadley joined the project because it was strongly recommended by her principal. She was glad to participate because she felt her mathematics background was weak.

Additionally, she welcomed the opportunity to learn best practices for teaching mathematics. Like many of the other teachers, Hadley was born and raised in the South. She attended a small church school and described her upbringing as strict and very conservative.

Jenna

Easily among the quietest teachers in the study, Jenna was raised in a racially homogeneous town of 5,000 people located in the Northeast. By her own admission she had not experienced much ethnoracial diversity until she entered university to pursue her undergraduate degree in teaching.

Jenna is a White first grade teacher at Juniper and she was a member of Cohort 3. She was a relatively new teacher, having taught for only three years. Like the other teachers at Juniper, she joined the project mainly because of her principal's recommendation and urging. Even so, Jenna did have her own goals for participating in NMD, and initially they aligned with her interest in mathematics.

In her mathematics autobiography, Jenna's memories of herself as a student include being good in mathematics and taking several advanced courses, including calculus, in high school and university. Jenna was probably the quietest member of her cohort although she seemed to find her voice when asked to contribute to the series of asynchronous discussions related to the two foci of the project.

Kendra

Kendra is a quiet, unassuming, African American woman who was in Cohort 1. She taught first grade at Farren Reese, where she had been for her entire eight-year career. Kendra was the youngest of six children, growing up in an African American community in a large Virginia city. She described her upbringing as one where excellence was expected and great family and community support was provided. Kendra joined NMD to learn more about mathematics.

She was motivated by her desire to better understand how her students thought about mathematics and how she could better teach it. In addition, she was interested in the role of culture and was intrigued by the idea of combining both math and culture in this project. She noted that as the only African American teacher on the first grade team at Farren Reese she was the one others turned to with their questions about culture in their classrooms.

Tasha

Tasha, an experienced, White, Cohort 1 teacher, taught kindergarten at Farren Reese. At the time the project began she had been teaching for sixteen years. Unlike many of the other teachers, she had experience across many grade levels, having taught kindergarten, first, and fourth grades. This experience gave her a longitudinal view of mathematics across the grades that was unique in her Cohort. Tasha joined NMD for the focus on culture and because it resonated with her own experiences in school.

Tasha had a difficult childhood. She quit high school in tenth grade to "survive." Eventually, she earned a GED (General Education Diploma or high-school equivalency certificate), worked at a day care, and went to college. She said that "teachers have been a guiding force in my life and I know that is why I became a teacher." Her passionate participation in project retreats made it clear that she was striving to be a guiding force for all of the children in her classroom as well.

DELVING INTO THE CASES

As can be seen from the above profiles, the nine case study teachers were a highly varied group. Analysis of data generated from some of the individuals who make up this select group of NMD participants offered poignant insights about why the project was successful. At the same time, data generated from other members of the group spotlighted a number of troubling findings that underscored the nature of the challenges future studies like NMD may well encounter. Thus, in the next two chapters we provide detailed profiles of the case study teachers with a focus on how each teacher grappled with the culture/mathematics focus of NMD.

Specifically, in chapter 4 we present an in-depth analysis of the experience of each primary case teacher who, to varying degrees, obtained from the project the hoped for outcomes. Then, in chapter 5 we detail the stories of each secondary case teacher. We explain how there was more variation in the reactions to and interpretations of NMD among the secondary case teachers. Still, their stories are especially critical because individually and collectively they emphasize both the urgency and the many complexities that invariably will accompany efforts to offer high-quality professional development for contemporary classroom teachers.

NOTES

1. Appendix A provides demographic information about the different schools and teachers who participated in the study.

2. High-tech industries, meat-processing plants, and agriculture are among the pull factors that have contributed to a shift in the demographic profile of the area. The population is highly diverse along various axes, including education and income levels. A distinctly liberal tinge has surfaced in the partisan politics within Crutchford-Whinton even as the surrounding rural areas preserve a deeply rooted conservatism that dates back many generations.

3. Year-round schedules have been implemented in many U.S. school districts to address classroom overcrowding and to make use of building facilities throughout the calendar year. For elementary schools that operate under a year-round schedule, teachers and students are organized into four separate groups commonly referred to as "tracks." Throughout the calendar year, three different tracks are in session (attending classes) while one is "tracked out" of session or on hiatus. Track-in sessions represent nine consecutive weeks of instruction, whereas track-out sessions last for three

consecutive weeks. Typically, all four tracks are out of session for the traditional U.S. winter holidays (e.g., Christmas, Hannukah, and Kwanzaa).

4. In DCY-2, cross-cohort bonding was introduced and facilitated through the cross-cohort dinners described in chapter 2. Then in DCY-3, the E-forum was introduced to promote three-way cross-cohort interactions. See chapter 2 and Marshall (2013) for a discussion of the NMD E-forum.

5. By the close of the project, a second teacher had earned a master's degree (in school leadership) and a third had begun a master's program.

6. Appendix B includes the NMD Mathematics Lesson Analysis Rubric.

Chapter 4

Dreaming a Bigger Dream

The Primary Case Teachers

The NMD project challenged teachers to grapple with serious issues related to mathematics teaching and learning in contemporary culturally diverse schools. Included among these issues was the connection between children's acquiring mathematics competencies and their future life chances (Moses and Cobb 2001; Paulos 1988). Teachers would be expected to examine sociocultural factors implicated in the early onset of mathematics underachievement among children from particular ethnoracial and economic backgrounds. Also NMD required teachers to evaluate their own mathematics acumen with a particular focus on their content knowledge and pedagogic skills.

In short, we offered an urgent and provocative challenge to teachers. It was one they soon recognized would necessitate multilevel reflection and introspection. Not all were receptive to the breadth of the challenge. Indeed, some resisted. Nevertheless, in time four embraced its full spectrum, causing them to stand out from among the rest. This chapter details their individual stories of professional development and personal growth. It begins with an overview of the retreat foci. Next we describe a retreat activity used early in the study to prepare the teachers for intellectual engagement. Finally, we conclude with the presentation of the four primary case study teachers: Johnetta, Giselle, Isabel, and Arlene.[1]

A BRIEF OVERVIEW OF THE NMD RETREATS

Each cohort of teachers attended a separate collection of two-day professional development retreats over the span of a project year. (See chapter 2 for a full discussion of the project retreats.) A challenge with the project retreats was to strike a balance between facilitating critical examination of tenets of CRP and exploring the principles and instructional practices aligned with SBM. In conjunction with these two foci, however, we worked to facilitate the teachers' abilities to observe and evaluate their own instructional practices.

Key elements of this latter effort included teachers viewing recorded segments of their own mathematics lessons as well as those of a same-status peer. Post-teaching reflection sessions were structured for teachers to focus on how they promoted (or hindered) their own students' conceptual understandings of the mathematics ideas they were trying to teach. We buttressed these activities with others designed to assess and build on the teachers' mathematics content knowledge. In the end, our analysis of interviews with the teachers revealed the various retreat activities, as well as the structured post-teaching reflection sessions, had been significant contributors to their growth during the project.

An especially important aspect of the project was to provide all of the teachers multiple opportunities to interrelate and exchange ideas with each other. Whether via retreats, intersessions, or mini-conferences, project activities typically were designed to promote multiple levels of interaction.[2] These different opportunities helped to solidify existing bonds as well as facilitate the development of new associations. However, it cannot be claimed that project activities facilitated a close collegial bond among the four primary case teachers. Indeed, as far as we could tell, there was minimal interaction between and among those four women outside of NMD. Still, we believe one experience, which was common to all the project teachers, may have prompted each of the four primary case teachers to begin engaging in critical examination (and arrive at a similar conclusion) about her stance on mathematics teaching, learning, and the impact of culture thereon.

CONTEMPLATING WHAT IT MEANS TO BE A PROFESSIONAL EDUCATOR

One experience, which occurred during the second retreat of each project year, seemed to set the stage for the level of intellectual engagement we sought among the teachers.[3] It was an activity we used to invite them to draw upon multiculturalism/CRP as a foundation to begin their own introspection and reflection on the quality of mathematics instruction they provide young students. Specifically, in preparation for this work the teachers participated in a series of conversations about the nature of professional identity.

They explored what professionalism means to them and then, drawing on their own experiences, they discussed how it does or does not manifest in schools and among classroom teachers. We invited them to identify ways the professional educator differs in a classroom setting from the nonprofessional or what we termed the "semiskilled technician." Most had never considered the implied duality of our terminology. Even so, they seemed intrigued by the opposing extremes and therefore engaged enthusiastically in discussion about the two.

We operationalized the concept *professional educator* by drawing directly on the work of Shulman (1998) and the three tenets of CRP (see chapter 1). That is, we proposed that due to their moral and ethical grounding, teachers who function in schools as professionals look for every possible way to facilitate success for their students. They have high expectations and they work to increase the possibility that each student will meet (or exceed) those expectations. The teachers themselves have a depth of knowledge of their content, and as such can promote deep conceptual learning among their students. Such teachers make deliberate efforts to *learn their students* (i.e., home circumstances, personal interests, learning challenges, and so on) while at the same time they work to better understand themselves as cultural beings (Chinn 2004; Gay 2002; Howard 2003; Ladson-Billings 1994).

In the latter case, we explained the notion of getting to know self (vis-à-vis the teaching role) as the process of becoming acutely aware of manners in which one's own ways of being (e.g., beliefs, attitudes, interaction patterns, and so on) may serve to disadvantage some students while at the same time offer unfair advantages to others. Further, we suggested that the professional exhibits a creative curiosity about teaching. (We expand on this concept in chapter 6.)

The notion of semiskilled technician, on the other hand, was described as a person who functions in the teaching role primarily as a maintainer of order and secondarily as a disseminator of information (Ladson-Billings 1994). We proposed that not uncommonly the information disseminated by such teachers is disconnected from the realities of students' lives outside of school. Multidimensional, "ill-structured" (Kitchener and King 1990), and unpredictable (Shulman 1998) teaching issues, which are common in the context of culturally diverse classrooms and schools, confound semiskilled technicians. As such, they are disinclined to take on complex intellectual tasks such as drawing on their students' cultures as part of instructional and curricular decision-making. We offered that the reluctance of semiskilled technicians to engage in such intellectual work might occur, in large part, because they are ill equipped to do so from both an intellectual and dispositional standpoint.

In relation to mathematics, we suggested that the semiskilled technician has limited and superficial knowledge of the mathematics she teaches. Instruction may be textbook dependent and minimally connected to children's real-world encounters with mathematics. Emphasis may commonly be placed on procedural knowledge of operations divorced from opportunities for youngsters to acquire deep, transferable understandings of elemental concepts (Ball, Lubienski, and Mewborn 2001; Hiebert and Carpenter 1992;

Kazemi and Stipek 2001). We concluded the discussion of the semiskilled technician by reading Haberman's (1991) germinal article describing the pedagogy of poverty.

At the close of our discussion about professionalism, each year we invited the teachers to take a personal stance on their own identities vis-à-vis the question, are you a professional educator or a semiskilled technician? Not surprisingly, all of the teachers rejected the semiskilled technician identity. Over time, however, we noticed the stance assumed by the four primary case teachers was substantively different from the others. Their teaching as well as their reflections on practice pointed to a different level of enactment of the identity of professional educator. These four women each exhibited an identity that, to varying degrees, was in alignment with the project's broad goals.

We have no reason to believe that the four primary case teachers directly assisted each other in arriving at their highly similar professional educator identity stances. For one thing these four joined the project in different years and they taught at different schools. Arlene and Johnetta were Cohort 1, Giselle was Cohort 2, and Isabel was Cohort 3. Three different project schools were represented among them.

Johnetta and Giselle, first and second grade teachers, respectively, were both from Rhine River; Arlene, a second grade teacher was from Farren Reese, and Isabel a first grade teacher was from Dawson Dome. With the exception of Johnetta and Giselle, it's our sense these four women scarcely knew each other. Johnetta and Arlene were in the same cohort, but were rarely required to work together. Since Isabel was in Cohort 3, it is unlikely she developed a close acquaintance with the other three, and vice versa. All were cordial during large group activities, but it is uncertain whether they knew each other. Indeed, it became known that at least two of them were definitely not friends.[4]

Our analyses of the various data sources revealed that individually they responded to the project foci (culture and mathematics) in a manner that in some respects was strikingly dissimilar to each other. This is to say although each was drawn to one particular focus of the study, over time they embraced both foci and arrived at a similar notion of the professional self. The experiences of these four women were notable because each adopted, and in her way enacted, the identity of *culturally relevant mathematics teacher.* We turn now to their stories.

JOHNETTA: A BIG PARADIGM SHIFT

Her voice is steady without a hint of timidity or reluctance. Johnetta has a lot to say and she is ready for her final one-on-one interview. The RA begins by saying, "Take a moment and reflect on your three years as a participant. Just tell me any of your thoughts and feelings." Johnetta responds.

> The first year was very trying, very taxing, very frustrating. Also it was a big paradigm shift for me because I hadn't been stretched mentally and academically in a long time. . . . The second year . . . I saw the impact of the program as far as with my math lessons. And one of those things that had the biggest impact was the buddy system. . . . That was a real trust building factor also because it was persons that you maybe work with, but you didn't open yourself up to [in terms of] that person having constructive criticisms about . . . you. . . . The third year it was probably not as stressful . . . . I felt a whole lot more relaxed.

She joined NMD at the urging of her principal, whom she describes as "a math person," but from the outset it was "the culture piece" that really interested Johnetta.

Over the span of the NMD project, Johnetta emerged as the teacher who most fully embraced and enacted the culturally relevant mathematics teacher identity. She challenged herself to grow in areas that she (and we as PIs) recognized were her least developed. At the same time Johnetta demonstrated the disposition and ability to re-orient her stance where necessary in those areas that were clearly her strengths.

Most impressive about Johnetta was her appreciation for the power of critical reflection on her teaching practices and self-image as a professional educator. One year after the classroom-based data collection for the project had concluded, she shared the following in a focus group interview:

> [In Crutchford-Whinton we] try new things and before we get a good handle on it, we abandon it and we try something else. Therefore you never really get to perfect what may have been great for student achievement. Whereas with this project, having been in it for three years, you can see the foundation being built and it was something you could build upon and continue to build upon and then it also could reshape your thinking as an educator. To me it's one of the most important things. And as I change as an educator, you can give me all the stuff and I can sit through, . . ., hundreds of workshops. But until I make a shift in my thinking and my teaching that will impact students, I [will have] just sat through another session.

Throughout her participation in the project, Johnetta consistently exhibited courage and commitment to work for, and realize gains from, critical reflection and introspection. The success of her efforts was apparent in the growth of her mathematics content knowledge, changes in her orientations to teaching, and more nuanced and thoughtful articulation of the complex impact of culture on the teaching-learning process.

Possibly more than any other teacher in the project, Johnetta understood the negative and potentially long-term effects of poor teaching in general, and low-quality mathematics instruction in particular. She knew how children's self-confidence could be diminished and destroyed by unbridled clashes between their lived culture outside of school and that which they encounter in school. Her empathy was palpable because she had "lived so much of that."

At forty-nine years old, Johnetta, an African American woman, was one of the more mature and more professionally experienced teachers in Cohort 1. She had been teaching just over twenty-five years when she joined NMD. Five of those years had been at her current school, Rhine River. Prior to her move to the Crutchford-Whinton area, however, Johnetta had taught for a much smaller school district in a rural area of the state she calls home.

> I was born and raised in the foothills. . . . My community consisted of family, family, and more family. The Purnells [her family of origin] were landowners, so they owned all the land that I could see that surrounded our home. . . . Our parents had no problem with [not] sparing the rod if it was needed to keep us in line. Therefore we were not problem children in school. . . . My formal education began in an all Black school. . . . My father had a saying . . . "What you learn no one can take from you, so learn all you can. They can beat you but they can't take what is in your head." These were some of the best days of my life.

In many respects Johnetta emerged as an exemplar for the type of growth we sought for all the teachers in NMD. This is so even though when compared to other teachers in the project, Johnetta's mathematics content knowledge was quite slight. This fact influenced her early reaction to NMD. Her scores (35.7 percent correct responses, with 42.8 percent for quality of explanations) on the TMQ were among the lowest.[5]

Although Johnetta displayed a positive and largely confident professional demeanor, it masked insecurities surrounding the slightness of her mathematics content knowledge. Near the end of the project we learned that "she had been apprehensive about NMD and wondered whether its *real* intention was to showcase the shallowness of the teachers' mathematics content knowledge. Further, she wondered if the plan was to spotlight African American teachers in particular as those most in need of professional development in mathematics" (Marshall and McCulloch 2010, p. 8).

Simply put, and by her own admission, Johnetta was not good at math. Reflecting on her decision to teach primary grades she shared, "I knew I had better be an elementary school teacher because if I had to [teach] math higher than third grade some child would be in trouble." But she had not always been so uncomfortable with math. Indeed, her recollection is that "math was easy to me during my grades one through seven." Drawing on vivid memories of herself as a math learner she detailed interactions with a Black male math teacher.

> [He] would pop us in the hand with a ruler if we missed math facts. . . . Even though we got popped I never felt humiliated or sadden, just accountable. The teachers rewarded you as well and we were expected to help the slow learner and for some reason we felt accountable for each other's learning.

The affirmation she experienced from teachers in the all-Black learning environment contrasted sharply with her experiences in the desegregated school in her small hometown. It was then that Johnetta encountered teachers who were both hostile and demoralizing to her as a math learner.

> Someone thought it would be a great ideal [*sic*] for Blacks and Whites to integrate so off [to] the white school I went kicking and screaming. It was awful. The teachers and White students were mean and rude. My learning took a downward spiral. I kept trying to keep my father's words of wisdom in my head but it was hard to keep your spirits up when you were being insulted and made to feel dumb. The few Blacks who were in the school banded together to support each other and . . . the church became our saving grace. . . . I feel my parents wanted us to respect the White teachers but they knew we were being treated as second-class citizens.

An especially vivid memory for Johnetta involves a White male teacher. "When I would ask him a question he would snarl and glare at me . . . so I drifted to the back of the room and decided I hated math and would never be good at it." Elaborating on the experience, she shares the following:

> It was based on who I was and what I look like. And he taught to the Caucasian males. So then that shift started [in my mind]. Well something's wrong with me that I can't get this. And he pretty much confirmed yes, something is wrong with you. So then my confidence in . . . learning math diminished a lot.

Recalling these details of her biography as a math learner seemed to represent a "breakthrough" for Johnetta. She later made connections to how she had been teaching mathematics to her young students.

> So then as a teacher, I don't think—, [let me rephrase that] I gave 100% but it wasn't 100% with confidence. It was "I'm doing this because the manual says this, this, this and this." After the NMD experience and I was able to put a face with all of that stuff that had been burned inside of me for years, [I realized] that what I do as an educator can impact the student years down the road. And then that just gave me the freedom to say there wasn't anything wrong with me. It was my learning style; the teacher wasn't sensitive to that and didn't care about

> that. So as a professional, it's my responsibility to the best of my ability not to do that to other students.

By her second year in the project, Johnetta was beginning to free herself from the web of math hatred in which she had been ensnared for so many years. She worked to increase her own competence and confidence in teaching mathematics. For example, analysis of her lessons in the first year revealed scarcely any evidence of CRP subskills[6] in her teaching; whereas in subsequent years she made abundant use of several skills, including *Illuminating Thinking*, *Relevance Making*, and even *Communalizing* (see chapter 6). From DCY-1 to DCY-3, we chronicled considerable growth for Johnetta in mathematics. At the same time, she exhibited important growth in her critical understanding of "the culture piece"—the area where she felt strongest.

Different from most of the other teachers in the project, Johnetta was eager to engage in discussions about the impacts of culture on the teaching-learning process. Early in DCY-2, she expressed ideas that reflected her efforts to make sense of this complex idea. In some respects her thinking was inconsistent with the focus of NMD, whereas in others it was on target. For example, when asked to explain how children's cultural backgrounds impact their learning, Johnetta responded:

> Students whose parents talk at them instead of to them, and [those for whom] exposure is not as great, I think sometimes have a tendency not to initially do as well. But [I'm] not saying they don't have the tools, but maybe it hasn't been tapped into.

This response includes allusion to a deficit viewpoint about the backgrounds/home lives of some children ("*Parents talk at . . . instead of to . . .*") while at the same time it includes acknowledgment that the substance of a student's knowledge base may be unknown to a teacher. ("*But maybe it hasn't been tapped into.*") Thus, in the midst of articulating her ideas about children's learning, she seemed to engage in a kind of self-correction that perhaps was grounded in a sensitivity borne out of her own experience with having been a learner whose background knowledge and very being had been devalued by teachers in schools.

An area where Johnetta exhibited substantial growth was surrounding the very concept of culture. In the focus group interview, she acknowledged that prior to NMD her notion of culture had been limited to *race*—specifically "African Americans and Caucasians." Over time, however, she began to reject (or at least no longer accept at face value) the longstanding U.S. dichotomy wherein historically, the population is bimodally divided into (and largely defined by) racial majority (White Americans) and racial minorities (all persons not classified as racially White).

When asked to share her concerns about drawing on students' cultural backgrounds when she teaches math, Johnetta said, "My biggest concern would be a misconception that I may have about their cultural background. Because my exposure has been limited pretty much to two races of people." Her expansion on this response demonstrates she had begun to understand culture more broadly—that she was moving beyond race as its sole component. She continued, "Probably the socioeconomic part, probably [my exposure is] very limited. So, a lot of it [what I would draw upon] would be built on what I've 'heard' or what I've read, or what I 'think' I know."

Another noteworthy finding regarding Johnetta's understanding of culture is related to the nature of the introspection in which she engaged as the project progressed. At the start of her second year she had already begun to think about culture beyond her personal experience of "Black/White." Asked what she would like to learn during the second year of the project, she replied:

> How the culture maybe of other ethnic groups like the Hispanic group or the Asian group, or the Hmong group, how their culture plays a part in the mathematical piece. So, I'd like to see or experience how it impacts what. . . . I do different as an educator [for those groups].

She expanded on the response in a manner that suggests she was acquiring a deeper understanding of the culture/teaching/learning connection. Johnetta acknowledged her own background may even provide particular advantages in her classroom for children who share her culture (i.e., African American children). She added, "As an African American teacher, based on my teaching style, I think the African American students possibly will be okay. It's the other group[s] that I have no personal experiences with."

It is noteworthy that Johnetta did not use definitive phrasing to express the academic advantage she seemed to imply could result for African American children in her class. Instead, her phrasing is tentative and measured (". . . *possibly will be okay* . . ."), which we believe suggests she was beginning to acquire a more critical understanding of the notion of culture as manifested in the realities of intragroup diversity.

Finally, unlike any other teacher in the project, Johnetta engaged in what can be termed speculative self-critique over the issue with which she had been most comfortable throughout the project. That was her fairness as a teacher. She noted,

> First, I think as an African American educator, my culture sometimes sets up some biases . . . in my classroom because—and I didn't realize this—most of us teach the way we were taught until we make that paradigm shift in our brain. One could honestly resent some of the Caucasian students who are just super at it because you're thinking that could have been me. . . . I had to check myself

> on that because it was my Caucasian boys who could pop it off in that math. It was almost like [I had] resentment, "Daggone it, how come they can get it and you [non-Caucasian students] can't get it?" . . . I had to address that and start to suppress that because that's not fair to all of the students. . . .
>
> I had to check myself and I still have to know who I am first and know what biases that possibly I bring to the classroom and [make sure] that it doesn't impact what I'm doing as an educator.

Johnetta had come a very long way from her start in NMD to the end of the project. She'd begun with the firm stance that she hated math, but by the end she described herself in the following manner:

> So, I view myself as a mathematical teacher that's passionate about teaching math because I do want them to have a[n] understanding of what it is we're doing and why we doing it. And that it's making them—or it's part of what they will become, which is lifelong learners.

GISELLE: I LOVE MATH SO, WHAT'S CULTURE GOT TO DO WITH IT?

When we first met Giselle she made it clear immediately how much she loved math. Whenever retreat discussions were focused on doing or teaching math she was engaged and confident. It was apparent she believed this to be the area in which she was more capable than her peers—and even sometimes more knowledgeable than the PIs.

However, we saw a very different Giselle whenever the conversation took a turn toward issues of culture. At those times she was quiet and hesitant to participate. This was an aspect of teaching and learning, and even of herself, with which she was visibly uncomfortable. The Giselle we encountered when Cohort 2 teachers joined NMD was quite different from the woman that left us at project's end.

Giselle describes herself as "a little White girl from a little white town." She was raised in a small upper middle class town in the Midwest. The town was known for its good schools and her parents chose to live in this town, with higher taxes, specifically because of the schools. Giselle had what she describes as an excellent K-12 education. According to her, this was especially the case for her mathematics instruction. She took Advanced Placement Calculus in high school and typically her math teachers were her favorite. After high school Giselle attended a small liberal arts college where she earned her undergraduate degree in elementary education. Later, upon moving to the Crutchford-Winton area, she completed a master's degree. When she joined NMD, she had been teaching at Rhine River for what to that point had been her entire six-year career. She was a second grade teacher.

The story of Giselle's growth during the NMD project is mainly centered on issues of culture. Even so, it is helpful to paint a picture of her mathematics instruction to explain the confidence she exhibited related to the mathematics issues in the project. A few years prior to joining NMD, Giselle had participated in a summer professional development activity that focused on the Cognitively Guided Instruction (CGI) project (Carpenter et al. 2000). This project highlights teaching practices that are very consistent with SBM principles and the related instructional practices we espoused through NMD. Giselle describes CGI as follows:

> It takes what children already know and you build upon that and you write problems to get children to solve them many different ways. . . . It's for all children regardless of how much experience you have, where you come from, what background you come from. All children can understand these problems. A lot of problems are written about a story, and so especially at the beginning of the year, we might take, like an Arthur story or a Clifford, or read a children's book. So we read through that so we all have the same experience with the book and we talk about it and things that we've been working on in class. . . . So the children have the same experience so that they have some knowledge already of what we're working with.

The CGI strategy requires students to work on particular types of problems structured in a manner appropriate for their individual developmental levels. This means not all students working on the same type of problem will be engaged with the same numbers in a given problem. Students are encouraged to use the tools and strategies that make sense to them. Likewise they are required to engage in whole class discussions centered on sharing strategies. Different from other approaches, in CGI the children drive most of the exchange.

Giselle was passionate about CGI and had even taken the initiative to provide training to other teachers at Rhine River. Soon after joining NMD she recognized parallels between aspects of the project's focus and CGI. Consequently, near the end of her first semester of participation in NMD she invited us to attend a national conference devoted to CGI. It had been scheduled at a nearby university for the following semester. We accepted her invitation. Later it was obvious that having done so served to affirm Giselle as a valued member of the NMD learning community.[7] Prior to that series of events, Giselle seemed to situate herself on the periphery of interaction.

For example, other teachers from her school related easily with us, the RAs, and with each other. Giselle, on the other hand, was scarcely part of their colleagueship and camaraderie. She rarely engaged in light conversation during session breaks or over lunch. More noticeable was that different from most of her Cohort 2 colleagues, she all but disengaged (unless explicitly asked a question) from discussions that focused on culture in the teaching-learning process. This contrasted sharply with her demeanor when it came to mathematics. At those times Giselle's excitement and interest were on abundant display.

Curiously, however, in the first pass of an analysis of Giselle's video recorded lessons we were a bit surprised and taken aback. This was because of the low number of tallies recorded for teaching practices that we knew to expect and that are consistent with CGI. Most notably was the absence of tallies aligned with the *Illuminating Thinking* (average of $n = 6.3$ across 15 lessons) category of our lesson analysis rubric. Still and all, the authenticity of the mathematics taught in her lessons was almost always rated as medium or high and the overall quality of the questions she asked was typically high as well.

To better understand the low number of tallies related to the use of high cognitive load interactions during teaching (i.e., *Illuminating Thinking*), we did a second pass analysis of her lessons. Upon completion we discovered many outstanding questions were being posed. Rather than Giselle's driving the conversation, however, her students were asking questions of each other. This sort of mathematical discourse was more sophisticated than we anticipated.

Indeed, our lesson analysis rubric had been designed to focus on teacher's actions and practices. As a result, it could not have captured instances within lessons where the students themselves were the initiators of the questions. In short, while we have no data support for our speculation, it is possible Giselle had developed a community of practice within her classroom that promoted and supported student generation of sophisticated questions and discussion about mathematics. This orientation to teaching was certainly highly consistent with the culturally relevant mathematics teacher identity we sought to promote.

Giselle's classroom practices were well aligned with SBM; however, at least one aspect of her practice contradicted both SBM and CRP principles. An RA expressed concerns that there were students in Giselle's class who were being overlooked during her teaching. Specifically, it appeared particular students were repeatedly being called on to share during whole class discussions more than others. On its face such activity clearly was inconsistent with the thrust of NMD. We suspect that at least one reason for this imbalance may have been Giselle's own ego and her self-image as a mathematics teacher. In this regard it is possible her desire to put her best foot forward caused her to also want to show off her best students on days her lessons were being recorded. On the other hand, we recognize that other, less benign issues could have been at play.

The simple fact is we had this very strong and confident second grade mathematics teacher who chose to join our long-term professional development project. In so doing, a challenge for her was the focus of the project was not mathematics exclusively. Rather, it also included critical explorations of the implications of tending to culture (both teachers' and students') on mathematics teaching and learning. When asked why she joined NMD, Giselle responded,

> I thought, well, this is just something else that I can learn from and grow as a teacher. And I like the math piece and I thought the culture piece would bring something that I'm missing because I'm just, you know, from a little suburb and I don't know much about other people. And I, you know, I teach here and I feel like I need to make sure that I'm meeting the needs of all my children.

Based on that response, it seemed she was receptive to and interested in exploring the issue of culture, yet her overall demeanor belied her words during that first year. Giselle was reluctant to discuss issues related to culture and when she did it was to offer ideas that clearly suggested she perceived mathematics as culturally neutral. She seemed reserved if not reluctant to engage about the topic broadly and even had trouble discussing her own culture. Later when reflecting on the project she shared, "I didn't even know my own culture. I remember that day when we examined our own culture and [my thought was] I don't have a culture. I don't have a race. I don't know. I have nothing."[8]

This perception of herself as unraced and cultureless carried over to her early teaching reflections. Whenever she was asked about the role of culture in her mathematics lessons she reacted with a nervous laugh. Then she would either say she didn't know, or she would evade the question by referring to the students' experiences within the context of the problems they were learning. One response proceeded thusly, "I guess some kids—you know, I don't have any this year that don't celebrate birthdays, but I guess if that was part of your culture you might be wondering why they get a card in the mail."

In the two years she participated in NMD we observed Giselle progress from being a young woman who saw herself as having "no culture" and avoiding discussions of the topic to one willing to participate in critical discussions within her cohort, school,[9] and on a national stage through participation in a professional presentation. This transition happened slowly, and mostly during the second year.

As the cohort teachers became more comfortable with each other they pushed each other during difficult cultural discussions in a way that was quite different from the mathematical discussions. Giselle was pushed by her peers to ask herself harder questions and to share her struggles more openly.

Her struggles and growth are evident in a reflection during her final semester in NMD. She wrote:

> I am adequately comfortable initiating conversation with colleagues about the impact of [culture on] the teaching-learning process [and] about [talking about] cultural diversity only because of the Dreamkeepers program. I would have never thought to initiate such a conversation before this because I wasn't as knowledgeable and I didn't realize the impact it had. I think now it is a conversation that needs to be had more often. I think the hard conversations that we had in Dreamkeepers led me to this level of comfort. It was hard to think about my own cultural identity because I didn't think I had one, but after we wrote our cultural autobiographies, I realized I did. . . . I think I am much more comfortable taking about race and culture with my students now.

An important event that foreshadowed the above reflection occurred after a national conference presentation session by two of the PIs and a select group of the Cohort 1 and 2 teachers. The presentation had been organized as a panel discussion. It opened with PI Marshall's providing an overview of the project goals and intervention design. This was followed with a description of the study's quantitative results wherein co-PI McCulloch presented a brief summary of the positive growth in mathematics understanding among both the teachers and their students. The presentation then turned to the teacher participants, who shared their stories and answered questions from the audience.

During a group lunch after the conference presentation the other teachers (many of them Giselle's Rhine River colleagues) commented on how far she had come. They recalled earlier, frustrating, conversations when Giselle had been reluctant to engage with them on issues of culture. Unanimously, they showered her with praise and expressed pride in how much she had grown. This interaction was one we reflected on as being an important watershed moment in Giselle's trajectory toward embracing the identity of a culturally relevant mathematics teacher.

Giselle had arrived at that point with lots of work remaining to be done; yet evidence was clear that she had made considerable progress along the multiple phases of change we identified among NMD teachers (see chapter 6) to claiming the identity of culturally relevant mathematics teacher. When invited to participate in a focus group conversation Giselle was more than happy to do so. Thoughts she shared at that interview offered insights into the continuing nature of her efforts to embrace the culturally relevant mathematics teacher identity. Therein Giselle reflected on what had been one of her more difficult teaching years.

> This year was by far beyond the roughest year I've ever had with parents, kids. . . . But I was like, these are your kids. You've got to form those relationships. And

> it was hard but I was like, no you need to walk the track with them again. You need to play on the playground with them. You need to form these relationships because they're not going to learn from you and you're not going to learn from them. . . . And these parents, I was like these parents are slackers. They don't come in. They don't do this. They don't read my newsletters. They don't whatever. And so then I [started] . . . calling parents and "well let's talk on the phone instead of [your] coming in." and I started writing positive letters. . . . I started forming relationships with the parents. . . . So I think the biggest thing that I had to do and maybe I probably wouldn't have done it if I didn't know how important it was to say, okay you need to get these kids and you need to have a relationship with them.

Giselle then told a story about a conversation she had with her students about whether or not it is important what color you are. According to her, that talk opened up a frank discussion about color and race with her second graders. She reflected on telling the other NMD teachers in her school, "I would never have had that conversation had I not done this. I would have never been comfortable to bring up 'let's talk about your races and where you come from'."

Thus, after two years of participation in NMD (and one year after its final data collection year) the "little White girl from a little white town" had come to recognize and understand the importance of the role of culture in mathematics teaching and learning. She had mustered enough courage to discuss it with her colleagues, and to think about how it affected her own teaching and interactions with parents. Significant for her, she even had come to recognize the importance of having critical conversations about it with her young students.

ISABEL: THE NOVICE DREAMKEEPER

Isabel came to NMD as a section chair from Dawson Dome and was strongly encouraged to participate in the project on behalf of her first grade team. She was charged to bring back relevant information to the school and to teach her fellow teachers what she learned with the ultimate goal of helping to improve student achievement. This proved to be a challenge because Isabel had a strong preference for teaching reading rather than mathematics. Also, she did not want to be the only person from her grade level to participate in the project. Despite this, she approached NMD with an open mind, eagerly engaging in the various activities with the goal of becoming a more effective mathematics teacher. As we examine her evolution within the project, we note how her early experiences influenced her thoughts about culture and the teaching of mathematics in the first grade classroom.

Isabel grew up in a small and predominately White community in a northeastern state. Her parents divorced when she was a child resulting in her spending the majority of her childhood in a single-parent low-economic home environment. Her schooling had little ethnoracial diversity. She is White as were all of her teachers and nearly all of her K-12 classmates. Isabel did not learn much about different cultures, as the school curriculum was not inclusive or multicultural.

In addition, she did not have the opportunity to interact with many people who were racially or ethnically different from her. One exception she recalls were a few Jewish friends in college. Isabel did not experience significant ethnoracial diversity until she relocated to the South to teach. She acknowledges that she had always thought of her experience as normal. "I did not feel like my experiences were different until I moved to the CWSD and began teaching. With such a diversity of cultures, races, religions, I really enjoy it!" However, these were not the only early experiences that proved to be influential in her life.

Isabel had numerous experiences with mathematics teachers, both positive and negative, that shaped her perceptions of mathematics. Overall she felt her K-12 teachers cared about her and her education. She described her relationship with teachers as caring: "I felt like teachers could relate to me because they lived in our small community and they had similar life experiences." Ultimately, this sense of similarity helped Isabel to feel more of a connection to her teachers. "As a student, that made me view teachers more than someone who is at school."

Although Isabel generally had positive experiences with schooling, she did not always have positive experiences with her mathematics teachers specifically. It seemed to her that the teachers emphasized procedural learning of mathematics; they required students to focus on learning the "rules" of solving mathematics problems. Also students were not allowed to question the rules. Because of this, by her own admission Isabel did not develop a conceptual understanding of mathematics. This made her very frustrated and subsequently it negatively impacted her sense of mathematics self-efficacy.

Things changed for Isabel when she enrolled in mathematics courses in college. There she learned why mathematics problems are solved in particular ways. In addition, she learned that teachers should provide students with conceptual and procedural explanations and that there are multiple ways to solve mathematics problems. This proved to be a turning point for Isabel and it affected her approach to teaching mathematics.

Isabel's lack of experience with ethnoracial diversity and her experiences with mathematics teachers shaped her views of how students learn mathematics as well as how teachers should teach mathematics. These experiences not only impacted her in the classroom but they also influenced her interactions within the project. As such, although her participation in NMD (as a member of Cohort 3) had been for just one year, Isabel was chosen as a primary case because she exhibited tremendous growth both inside and outside of the classroom.

She exhibited high levels of CRP and SBM in her classroom practices in conjunction with her participation in NMD as indicated by our use of the lesson analysis rubric. Specifically, we saw various instances of *Illuminating Thinking*, *Learning Connecting*, *Relevance Making,* and *Cultural Connecting* (see chapter 6).

In addition, her initial score on the TMQ was 62/100, and she offered superficial explanations for how she arrived at solutions to the problems. This contrasted with her scores by the end of the project where she earned 78/100 and offered explanations that were more conceptually grounded. In short, Isabel began the project with little confidence in her mathematics ability, yet she ended it wanting to learn more about mathematics as well as how to better combine mathematics and culture.

Through participation in NMD, Isabel changed her mathematics classroom practices to focus more on student learning. This meant she made a shift from a teacher-centered classroom to one that was more student centered. Likewise, her instruction emphasized both procedural fluency and conceptual understanding. Isabel described her new approach to teaching mathematics thusly, "I enjoy making mathematics real to them [my students]. I try to make examples, lessons and homework that is real for them and fun. I also make sure their homework is work that they can do independently." She also changed the manner in which she interacted with her students.

The children in Isabel's class were more engaged with the mathematics, which meant they were encouraged to use more mathematical language. For her part, Isabel incorporated a more active, high-cognitive loading questioning style with her students. She describes this change:

> I used to always do a whole group and then go back to your seat and do independent work. I have them talking more. And before I wouldn't [walk around the classroom]. I would kind of listen to them but I wouldn't ask more extensive questions. Now I'm like why do you think that? How did you get that? So that way they have to talk it.

Similarly, Isabel discussed how she began to listen more to her students to understand their needs. She described how the design of her teaching changed based upon the knowledge she gained about her students. This information she used instead of following a preset generic teaching schedule and plan. Also Isabel detailed how she began incorporating peer-teaching techniques. She explained:

> When I am teaching a lesson and see that some of the students are frustrated or not understanding the concept, I will revisit the concept with them and try to explain it another way. I also have asked students who have a good understanding of the concept to work with the student. I will give the students a challenge with their work daily. "Can you figure out another way to solve the problem?" or I want to see some Level 4 thinking . . . trick me! They love this challenge and it has really worked.

Once Isabel changed her approach to teaching, other teachers at her school began to notice changes in her students. Specifically, her first grade team noticed that Isabel's students made tremendous growth in mathematics over the school year: "I was so excited because they [my first grade team] were like your children made such growth; I cannot believe it." Because of this, Isabel was motivated to push for more continuity in the teaching of mathematics across grade levels. She began to think more broadly and considered how to best prepare K-2 students.

These reflections resulted in Isabel's beginning to see the need for cross-grade collaborations. Particularly she sensed an urgency to make connections within the kindergarten, first, and second grade curricula: "With math especially, we need to sit down across grade levels and talk to each other." She shared her concern with the school administration and was hoping that cross-grade collaborations could become the norm at her school. Until that time, however, Isabel was determined to continue to learn and grow professionally, and in turn provide the best preparation for her own students: "I'm always like how can I better myself therefore how can I better my students and make them more prepared for second grade." As she was becoming more proficient in teaching mathematics, Isabel also began to consider the role of culture in students' learning.

Specifically, she began to understand the connection between students' family lives and their schooling experiences. This she followed with making a concerted effort to get families involved in students' learning. She shared her emerging viewpoint: "Building a relationship with families is the number one thing." One way she tried to build relationships was by creating a heritage unit wherein students traced their lineages and countries of origin. This project was extensive and required family involvement. Isabel was very proud of the heritage project and felt it helped to create a connection among her, the students, and their families. She viewed the connection between the home, teaching, and learning as essential. She explained:

> So being able to look at my class and being able to try to figure out okay what is going to be the best way to instruct them, not necessarily for me but for their learning. Not necessarily what's the easiest for me. And to put into factors of

their home life and to get to know the families more and to find out where they're coming from so that I could better be able to teach the children.

By the end of the project, Isabel recognized how much she had grown within her own classroom. She was now using a variety of teaching strategies that allowed her to focus on student learning.

Participating in NMD helped Isabel to develop a more positive identity as a mathematics teacher and become more comfortable teaching mathematics. She saw herself as a mathematics teacher cognizant of the impact that her students' cultural backgrounds can have on the teaching-learning process. Although Isabel was pleased with the progress she had made while participating in the project, she realized there was much more she needed and wanted to learn about mathematics and student learning. After participating in NMD, Isabel shared that she was looking forward to enrolling in other mathematics professional development programs so that she could become an even better mathematics teacher.

ARLENE: AN EXTREME MAKEOVER

It was the final presentation Arlene would make to a group of her fellow NMD teachers. As a member of Cohort 1, the stipulation was that she had to "give back" to others in the project. This time Arlene's presentation was before a group of Cohort 2 teachers. Arlene began by sharing that she had come a long way over the three years she'd been part of NMD. For her final presentation she prepared to share just how meaningful the experience had been.

The opening slide revealed the title of her presentation: "NMD Extreme Makeover," and the next two slides showed photos of Arlene. The first was a disheveled image of her under the headline "Before NMD"; the second revealed a polished professional photo with the headline "After NMD." The images had been displayed for comic effect, but to Arlene they communicated so much more. Turning serious Arlene detailed how her orientation toward teaching mathematics and her understanding of culture on the teaching-learning process had changed over the three-year span. According to Arlene she had been "made over" both professionally and personally by the experience of NMD.

Arlene was born in a small town about 75 miles outside of Crutchfield-Whinton. She completed an undergraduate degree in Psychology at State University and later earned a teaching license at a nearby four-year institution.

These accomplishments distinguished her as the first grandchild on either side of her family to graduate from college.

Arlene describes herself as having been an "economically challenged" child because she received reduced price lunches early in her schooling career. Her mother worked two jobs to make ends meet. Over time things did change in her family for the better leaving Arlene with fond reveries of "moving on up."[10] Mom remarried, eventually gave birth to Arlene's two younger siblings, and they became a "two-income family [that] no longer needed to struggle . . . to make ends meet."

Her cultural autobiography begins "I am an American Southerner of European decent [*sic*]. The majority of my ancestors were from England, Scotland, and Ireland." She writes that memories of her early teachers are foggy, yet this claim is belied by her descriptions of a few who clearly made important impressions. For example, Arlene credits Mrs. Fauntroy, her favorite elementary school teacher, with being a major influence on why she decided to become a teacher. Mrs. Fauntroy, Arlene recalls, "cared a lot for me and all her students."

Other precollegiate school interactions that made an impression include learning Spanish from a teacher "of Anglo American background" and the twelfth grade math teacher, Mrs. Smith, who was a favorite. "All through elementary school my teachers were White females. I do remember having some African American teaching assistants. One in particular, Ms. Harrison called me motor mouth." Other sharp memories for Arlene include a seventh grade English teacher. "My least favorite teacher . . . was Mr. Washington [who] was African American and my first male teacher that wasn't a gym teacher. He was crazy about diagramming sentences."

It is striking how easily Arlene names ethnicity when describing herself and her Spanish teacher. During her extreme makeover presentation, Arlene recalls that during her first year in NMD she did not consider herself "ethnic." Instead, for Arlene, the term "ethnic" had always been used in reference to "the minorities"—African Americans and the growing population of Latinas/os. Three years of exploring issues surrounding culture and the teaching-learning process, however, had brought Arlene to a place where she now recognized that she, too, has an ethnoracial classification. During her final presentation it is one with which she clearly and proudly identified.

Other indications of Arlene's "makeover" vis-à-vis cultural understandings were evident in her responses to one-on-one interview questions and prompts. Reflecting on the overall experience of the project, Arlene shared the following:

> After participating in NMD, I do feel much more comfortable talking about cultural issues. For the first time in my life, one of my best friends is from a

> different cultural group than my own. I attribute this new level of comfort with my own culture and with cultures different than my own to the candid conversations we had in the safe environment of NMD. I just hope that I can carry over my new comfort level and deeper understanding of the role of culture in both the teacher and students' lives to my new school.

Arlene's growth vis-à-vis the culture focus of NMD is apparent when we contrast her end-of-project thoughts about culture with those she expressed during an interview at the close of her first year in the project. Asked to share what she'd like to get out of NMD as she entered the second year, Arlene's response focused on culture but in a manner that suggested she was searching for prescriptions to enhance how she works with other people's children. She noted,

> I think I would like to learn maybe more resources that are available to us—or just some more ways to incorporate it. I mean, you know, there's certain ways in the book that we read,[11] that other teachers have done. But I think it would be worthwhile as a group to (inaudible) talk about, have a plan, that kind of thing, about what we can really do in [Crutchford-Whinton]. . . . You know, what resources do we have? And so I would like for us to get specific with our area.

As one of the project teachers who demonstrated one of the highest scores on the TMQ, Arlene shared that NMD had brought greater clarity to her understanding of various concepts in mathematics—ones she'd taught for years, but on some levels had never fully understood. Reflecting on her experience as a math learner Arlene wrote, "Mrs. Smith, . . . made math make sense for me."

But these fond memories notwithstanding, Arlene recalls her teachers often did not teach to her learning style. In relation to mathematics, Arlene shared, "I was often confused in math." Then, drawing directly on terminology incorporated in all NMD retreats and activities, Arlene detailed, "I think that procedural teaching was just common practice at the time. They didn't teach for conceptual learning, which is what I needed."

Different from and compared to most of the other teachers in the project (across cohorts, but particularly among her own Cohort 1 peers) Arlene's background in mathematics was quite impressive. She had taken Calculus in college. Yet detailing her own mathematics story, she revealed that she had lacked clarity about mathematics concepts. In short, Arlene had always been unsure about herself in terms of mathematics. "I remember being confused about the 'whys' of math; I just followed the rules that were shown to me and memorized the appropriate algorithms and facts." By the time she was in her teacher licensure program her "eyes began to open up to math." She expands:

and personally, that in different yet equally profound ways propelled her to want to learn more about the complexities of teaching mathematics. More importantly, each of these women exhibited the curiosity to learn about both facets of the project, and the courage to challenge herself when the proverbial "going got rough."

Through participation in NMD, these four primary grade teachers, each in her own way, dared to examine what it means to be a professional educator in contemporary culturally diverse schools. In the process, each gained insights into the teacher she was, and started herself on a journey to the teacher she wants to be.

In the next chapter, we explore our secondary case teachers. They too, in their own ways, gained from participation in the project. Yet different from our primary cases, the experiences of our next five teachers were at once celebratory and confounding.

NOTES

1. Details for each case were drawn from multiple data sources, including cultural and mathematics autobiographies, one-on-one interviews, post-teaching reflection sessions, lesson analysis rubrics, field notes, and post-NMD focus groups. Three focus group interviews (with different participants in each) were conducted with select teachers one year after the data collection period for the study had concluded. The stipulation for invitation to participate in a focus group was that each teacher had to be identified (through independent selection) as having made *noteworthy professional gains* as a result of her participation in the NMD project by at least two different groups of evaluators (i.e., PIs, RAs, Principals). Two of the primary case teachers were selected to participate in a focus group interview.

2. This included within grade, within school, cross-grade, and cross-school. In addition to these levels of interaction, teachers were encouraged to engage in cross-cohort interaction. See Marshall (2013) for a discussion of the NMD Moodle E-forum activity designed to promote multilevel interactions among the teachers.

3. Teachers who participated in the project for multiple years (Cohorts 1 and 2) engaged in this activity during the second retreat of their first year only.

4. Giselle and Johnetta taught at the same school, but theirs was a strained collegial relationship, which one of them made known during a cross-cohort intersession meeting. The two were assigned to work together as part of a cross-cohort/cross-grade activity. Prior to the start of the activity in a private exchange one asked, "Do I have to be in a group with her?" PI Marshall pointed out that other teachers would be in the group as well. Ultimately, neither Giselle nor Johnetta displayed overt antagonism toward each other. Indeed, months later both participated in a national conference presentation panel discussion about the NMD project.

5. See Appendix C for sample items of the TMQ. Neither Johnetta nor any of the other teachers were made aware of their scores on the TMQ until near the very end of

the project. Even then, we reported TMQ findings in the aggregate. We believe that to have revealed scores sooner would almost certainly have been demoralizing for many of the teachers.

6. We explain this concept in chapter 6.

7. PI Marshall submitted a proposal to present at the conference, which was subsequently accepted by the organizers. In the presentation, we (Marshall, DeCuir-Gunby and a project RA) explored connections between the broad focus of NMD and CGI. Following up as we did on Giselle's invitation helped to strengthen her bonds with the NMD project.

8. This sense of self as unraced and cultureless is a common phenomenon among many Whites. Various scholars trace its source to the nature of White racial identity development, complex denials of race and racism in White socialization, as well as White racial privilege (see e.g., Helms 1993; Howard 2006; Matias and DiAngelo 2013; McIntosh 1990).

9. When the project concluded, the NMD teachers from Rhine River (including Giselle) arranged for PI Marshall to give a presentation on CRP to the entire teaching staff as part of the school's professional development goals.

10. Here, Arlene references *The Jeffersons,* a mid-1970s television series about a fictional African American family who acquired wealth through the success of the patriarch's dry cleaning business. The phrase "moving on up" is borrowed from the program's opening theme song. It is a reference to the family's upward mobility and their subsequent physical relocation. As a direct result of their financial success, the Jeffersons moved from their working class neighborhood in Queens to the east side of Manhattan. The setting for most episodes in the series was their high-rise condominium described in the theme song as a "deluxe apartment in the sky."

11. All of the teachers were given a copy of Ladson-Billings' award winning book *The Dreamkeepers: Successful Teachers of African American Children.* Cohort 1 and Cohort 2 teachers read, discussed, and analyzed the book over the span of three and two years respectively, whereas Cohort 3 teachers explored its substance for just one year.

Chapter 5

A Theory/Practice Chasm

The Secondary Case Teachers

The overwhelming majority of teachers across all three cohorts gained from their experiences in NMD. This was confirmed by the findings from our various formal data sources, as well as feedback we received from principals and the teachers themselves. It seems by and large, NMD was superior to other professional development experiences offered through CWSD. Consequently, the accounts of professional renewal and personal discovery for our primary case teachers were affirming on multiple levels. Yet it soon became obvious that theirs were not the only stories to emerge from NMD.

Rather, we stumbled upon an especially confounding case, a cautionary tale as it were, of a Cohort 2 teacher. Then, with subsequent data mining, others began to surface. There were five in all. Each offered distinct, and in some respects troubling, contradictions to the descriptions of breakthrough and connection that had typified our four primary case teachers. In this chapter we present these five different and compelling stories.

VARIATIONS ON THE DREAM: EXPLORING THE SECONDARY CASES

The stories of the five teachers we present in this chapter represent our secondary or *negative* cases. Broadly defined, a "negative case" represents a feature of the data that contradicts, does not fit, or is in some manner inconsistent with the pattern of findings in a given study in whatever way that pattern has been defined. Investigating negative cases is a common technique used in order to help establish credibility by enhancing the quality and rigor of analysis (Lincoln and Guba 1985). Our choice to present these five cases lets us explore variations in the experiences of the teachers who participated in our study. Through these cases we further illustrate the messiness of real-world research, a goal of this book as described in chapter 2. At the same time, acknowledging negative cases can help to broaden ideas, change perspectives, or question conceptualizations (Patton 1999). For us, these secondary cases helped clarify a perceived theory/practice chasm that for a number of the teachers was troublesome and distracting.

Brodsky (2008) describes that "negative cases are not a rare occurrence, but rather are a natural part of any study" (p. 4). Further, negative cases in qualitative research analysis may emerge from a deliberate or spontaneous search of the data. Our first negative case emerged spontaneously from our disbelief and curiosity about inconsistencies between our perceptions of one teacher, and what we took initially to be mere anecdotal impressions of her classroom practices as reported by one project RA. However, subsequent cases emerged from our deliberate and careful searches of our data for unusual situations vis-à-vis the broad pattern we had expected to find among teachers in our study. Briefly, this pattern was one in which teachers' engagement in the retreats was expected to be purposeful and reflective. Teachers were challenged to struggle with meaningful mathematics, and in turn, increase their own mathematics knowledge.

Our expectation was to see changes in the teachers' mathematics pedagogy. Hence, retreat experiences were structured to provoke changes toward more student-centered teaching. This orientation we expected to couple with a more nuanced understanding of (appreciation for) the impact of elements of culture in schooling in general, and mathematics teaching and learning in particular. Therefore, to identify our secondary cases, we homed in on inconsistencies among our perceptions of teachers based on our interactions with them during retreat sessions, prominent themes in their post-teaching reflection sessions, and outcomes (tallies recorded) in analyses of their actual mathematics lessons.

We hasten to note that while we have used the term "negative case" interchangeably with "secondary case," the experiences of this group of five teachers cannot be characterized uniformly as unfavorable, bad, or even wholly unsuccessful. On the contrary, all five teachers reported they had gained from participating in the NMD project. Four of them enthusiastically embraced both its culture and mathematics foci, albeit to different degrees. Each seemed to embrace fully the project's goals. As a result, for us the inconsistencies arose in part from our own inability to offer strong confirmation (through agreement between our qualitative and quantitative measures) of the *extent of benefit* these teachers gained from NMD. We believe this may speak more to the shortcomings in our measures than to the veracity of the teachers' self-assessments.

Only one teacher from among these five stands out as having been explicitly resistant to aspects of the project. Moreover, the resistance we uncovered was ensconced in layering that, barring deep analysis, could easily have remained undetected. This is because our preliminary analysis, which focused more on feedback from RAs and our interactions with the teachers during retreat sessions, proved inadequate to reveal how aspects of the project had provoked personal discomfort among a small minority of the teachers. Thus, this lone "resisting" teacher was included among our secondary cases because of the extreme viewpoints she eventually did express about aspects of her experience.

PRACTICE, THEORY, AND AN INADVERTENT GAP FOR SOME TEACHERS

A significant finding in our data analysis was presented when we found that numerous teachers struggled to mesh techniques they were learning to associate with SBM with ideas being explored as part of CRP. Particularly niggling was although their own basic understandings of mathematics content had improved, many teachers simply were not drawing on children's cultural backgrounds in the context of mathematics lessons. Consequently, it became evident that even teachers who expressed strong commitment to integrate CRP and SBM were at a loss about how to make that happen.

Because of this, for some teachers the NMD experience may have inadvertently reinforced beliefs about an inevitable theory/practice gap between university-based teacher education/professional development and the real world of schools (Shulman 1998). It was apparent something would need to change to lessen the perception that project retreats were the fluff of "theory" about CRP/SBM connections, a fluff some teachers perceived could not be brought to fruition in the real-world "practice" of actual lessons. We therefore created an activity for the teachers to complete during a retreat where they watched and analyzed a segment of their own teaching, and then wrote a reflective analysis of how they had (or could have) incorporated CRP into the lesson.

Admittedly, a challenge of formulating that particular activity was that it required us to operationalize components of CRP that prior to that time we had resisted. A cynical interpretation of our actions might suggest that up to that point we were beholden to orthodoxies and prevailing axioms in the professional literature surrounding cultural relevance as a pedagogical stance rather than a set of discrete teaching techniques (Ladson-Billings 1994, 1995, 2006). Such a claim might also imply that our initial resistance to operationalize CRP inadvertently aggravated the theory/practice chasm some teachers associate with professional development in general, and critical multiculturalism in particular.

It is true that we did feel a pressure to operationalize both CRP and SBM, and we explore the reasons for this in the final chapter. For now, however, suffice it to say that although a unifying theme in the experience of the five secondary case teachers was their struggle with the culture/mathematics connection, ultimately, they are distinguished for many different reasons. On the one hand, aspects of their experience left us confident that the goals of the project had been realized; whereas on the other hand various contradictions contained in their individual experiences left us humbled by the sheer enormity of the research task we had undertaken. Our secondary case teachers are Aisha, Tasha, Kendra, Hadley, and Jenna. In the next sections we present their stories.

AISHA: REQUISITE DISPOSITIONS AS THE RIGHT STUFF, OR NOT

She was a thirty-nine-year-old African American woman with varied experiences as an educator. When Aisha joined NMD she had been working in the field of education for nearly fourteen years. She held a master's degree in adult education with a focus on diversity training. Among her professional experiences were classroom teaching and instructional coaching. In the latter capacity Aisha had been responsible for presenting professional workshops to teachers on the topic of integrating diversity into literacy instruction. When she joined NMD she was a kindergarten teacher at Rhine River and a member of Cohort 2.

Aisha's family of origin was solidly middle class. Both her college-educated parents worked in higher education well into Aisha's adulthood. She recalls as a child having had substantive interactions with persons from diverse backgrounds. Her early exposure to ethnoracial diversity had a significant effect on her professional identity. She notes,

> I know that I [am] the teacher I am because of the natural affinity that I have for people of other cultures and my love of learning! My mother has traveled to every continent except Antarctica. She has an affinity for the international students at [university name]. I never knew who she was going to bring home. We have had students from India, Africa, and Hungary at our home on several occasions.

Although we introduce her as a secondary/negative case for analysis, in many respects Aisha was a definite positive asset to the NMD project.

She engaged readily and often in all activities (both small and large group) and seemed genuinely focused on enhancing her teaching skills. Likewise, she was effective and instrumental in challenging other teachers in the project to think critically about issues.

The consensus perception among the PIs as well as the project RAs was that Aisha was among the strongest and most knowledgeable of the teachers (across cohorts) in relation to the CRP focus of the study. Owing to her broad knowledge, she was one of the teachers we invited to share personal reflections on her experiences in the project during our third (and final) year recruitment efforts (see chapter 2). Also, Aisha was invited to participate in the national conference presentation that featured a panel of teachers from the project. She was an engaging speaker with a natural flair for the comical.

Irrespective of the topic, whether in large group discussions or small group activities, Aisha did not hesitate to engage. Always, she could be counted on to share thoughtful ideas and opinions about the importance of teachers' attending to cultural diversity through their teaching strategies, instructional resources, and general professional demeanor. The ease and care with which she engaged others was particularly on display during a retreat activity and discussion about ethnoracial identity development. Another African American teacher and fellow Rhine River colleague shared personal challenges in dealing with her own identity as an African American woman. Compassionately, Aisha urged the teacher to consider the impact such a struggle could present for young African American students.

Aisha seemed to really understand the complexity of multicultural education in general, and CRP as an equity pedagogy in particular. Using her own schooling experiences as an example, below she exhibited awareness of the multidimensionality of diversity in schools as well as an understanding that the challenge of CRP extends beyond issues of race.

> Unfortunately, I had a white math teacher who proved to be not only a racist, but an elitist too! . . . Thank God that I had [names two other teachers] in the same year to help me stay positive. . . . For the most part I enjoyed all of the teachers and I felt like they related to me because of my middle class background, but not necessarily my race.

Likewise, in her professional self-image, taken from her mathematics autobiography essay, Aisha seemed to recognize that adopting CRP demanded she attend to the needs of all her students.

> As a teacher of kindergarten students, I feel confident in my abilities as a math teacher. . . . The repetition of reviewing concepts in different ways allows all of my student's opportunities to learn! . . . I am a good math teacher because

> I don't teach to the middle. The students who are struggling can ask questions without me thinking something is wrong with them. Many times the math teachers in the past had the attitude that math is exact! There is only one answer and why can't you get it right!

Through our multiple interactions with Aisha, we held a very firm impression that she exhibited requisite dispositions (if not actual skills) to merge tenets of CRP with SBM. Our broad perception was that the dispositions Aisha exhibited manifested through her openness, if not eagerness, to examine critical issues including implications of teachers' expectations for children of color. She readily articulated an understanding of the difference between mere awareness of cultural diversity versus adopting a pedagogical orientation with potential to result in substantive positive learning outcomes for students in schools.

Drawing on her cultural autobiography, we note Aisha's unprompted attention to exploring controversial issues and professional beliefs/practice congruence. She stated,

> Educators will continue to do business as usual [in their teaching] until they are willing to have those uncomfortable discussions. Discussing diversity as it relates to teaching and learning is just the first step. Real change will not occur until the entire county or even the school began [*sic*] to apply what they learn to their everyday habits and plans.

Contradictions were presented when we analyzed lessons and post-teaching reflection session data sources. They revealed a much different image of Aisha from the one we gained through interactions with her during retreats and project activities. The biggest inconsistency was that the teaching acts we saw (and later analyzed) did not mesh with the abundant knowledge of and appreciation for CRP she consistently presented in retreat and other contexts. Most striking was the lack of evidence of Aisha's ability to reflect critically and thoughtfully on the reasons behind her pedagogy.[1]

Aisha readily acknowledged that it was not concerns about culture and addressing diversity in teaching and learning that had been the appeal for her in joining NMD. Instead, it was her interest in enhancing her knowledge of how children learn mathematics that would be the challenge. A big strength was that at the start of the project Aisha recognized her skills as a math teacher needed work. In recognizing this weakness it seemed Aisha was demonstrating the ability to self-critique. By the end of the project, her perception was that she had gained in that area. When asked to reflect on her gains she noted:

> I had written that I wanted to understand how complicated number sense really is. And I learned that through NMD because I always thought okay well you count and you see it's one to one. I never understood what's so hard about that.

> Even though math is not my favorite subject, I thought that kindergarten math is not brain surgery.
>
> But through participating in the project that focused on K-2, it really helped me understand what number sense is really like and how long it actually takes to master that. And that's really powerful because if kids don't have number sense and they can't be successful and go on through the rest of their years in math, they'll struggle. Which makes me think I didn't really have good number sense.

Her thoughts about actual changes in her students in conjunction with her participation in NMD were as follows:

> I think just practicing the questioning for me has really helped them with the answers. So I've learned and I think their growth is through my growth especially when [project co-PI] taught us how to ask the questions without value judgment and things like that. That's very powerful because it's really hard for teachers not to say "great" or calling the kids to have to answer but working with her really helped us understand the power of questioning.

Thus, for us, the conundrum of Aisha was that her articulate insights in the context of retreats as well as the thoughtfulness of her responses in one-on-one interviews contrasted sharply with the pedagogy she enacted in her classroom. The depth of her knowledge about CRP during retreats was wholly inconsistent with the nature of the analysis she offered about her lessons as part of post-teaching reflections on her pedagogy.

Over the span of two years, researcher notes recorded as part of our analysis of twelve different lessons taught by Aisha paint a near-opposite image of her when contrasted with what we experienced in the retreat context. Box 5.1 presents excerpts from researcher notes in the first and second years of Aisha's participation in the project.

Analyses of her lessons revealed virtually no instances of CRP subskills, and there are multiple expressions of concern expressed about the substance of the mathematics itself. Even in the area where she perceived she had grown most (questioning strategies) the data suggests something quite different.

Every lesson she taught reveals that overall the types of questions Aisha posed to students were "low quality." This was defined in our lesson analysis rubric as "queries that require students to exhibit memory of previously learned materials by recalling facts, basic concepts, and answers." In short, we identified substantial evidence to suggest that the very clear and well-articulated understandings of CRP Aisha was able to contribute to discussions in project retreats did not manifest in her actual classroom practices. Likewise, her orientations to teaching mathematics were in some respects opposite to what was being advanced in the project retreats and what she seemed to think (vis-à-vis her reflections) she was actually doing in her lessons.

Box 5.1 Notes on Aisha's Mathematics Lessons

FIRST-YEAR ANALYSIS NOTES

- Limited amounts of language matching
- If the students are not answering her specific question she does not want to hear it
- Uses equity sticks (but only picked on advance students)
- Again VERY procedural
- Too concerned with the rules of the game/activity and not enough about the math content
- She uses a lot of worksheets; they include cutting and coloring
- She is very concerned with the students following her specific rules more than the math content

SECOND-YEAR ANALYSIS NOTES

- Very procedural
- Asks [illuminating thinking] question[s] but then answers them herself
- The students do a lot of small activities. They take about five to ten minutes. However . . . does not explain the connections between the activities.
- Teaching is always interrupted by her correct[ing] the students' behavior

Ultimately, the story of Aisha is one of the more baffling of our secondary cases. In part, this is owing to the fact that the issues we found troublesome in our reflections on Aisha's experience inadvertently pointed to a number of our own missteps as researchers vis-à-vis the design and implementation of the data collection protocols for the study. Thus, while baffling, the lessons we learned from our work with Aisha are among those that offer us the greatest insights into the implications NMD presents for future research. We elaborate on these lessons in chapter 6.

TASHA: I THINK I CAN, I THINK I CAN . . .

> Just to make sure that I am as culturally aware as I need to be . . . I really want to make a difference in all my children . . . and it's going to make me a better teacher and help them learn more. . . . That was my goal.

That was Tasha's reason for joining the NMD project. Tasha, a Cohort 1 teacher at Farren Reese, was passionate about NMD and becoming the best teacher she could be. She cared deeply about her students and as such we saw her grapple with both constructs of the NMD project. Tasha explained, "I feel like I pride myself on the fact that I am really loving . . . but I also feel like I know I could do a better job. And I want to make sure I do that." Tasha is a confounding case because though we saw significant changes in her professional stance toward both culture and mathematics, the changes in her beliefs resulting from what she had apparently learned often did not play out successfully in her classroom.

When Tasha joined NMD she did not hesitate to let people know she felt mathematics was her weakest area as a teacher. She was not comfortable with the content and was not used to thinking about her students' thinking. Even so, she embraced much of what she was learning in the retreats and immediately began to try to change her mathematics pedagogy.

She implemented open-ended tasks, asked her students thoughtful questions, and made a concerted effort to make sure that the onus of responsibility to generate mathematical knowledge in her classroom was not only held by her as teacher, but also by her students as learners. These were significant and risky changes for someone admittedly not confident with the mathematics and whose lessons had been highly teacher centered for sixteen years. Tasha noted that with these changes her students were more engaged and they appeared to be doing better on assessments.

We were surprised to find that when Tasha engaged in post-teaching reflections about her lessons typically, she was unable to identify the learning goals. She had difficulty discussing what her students had learned and how she knew they had learned it. Additionally, she was often concerned about what to do next because she felt the lesson had not gone well. These difficulties may have been due to her discomfort with the mathematics content, or maybe her inexperience in implementing the particular types of lessons. Regardless, Tasha did not give up. She often used the reflection time with her buddy discussing how to improve, asking for very pointed help, and noting things that had gone well in her buddy's lesson. Commonly, items in her buddy's lessons were those she planned to try in the future. Unfortunately, at the end of three years, Tasha was having the same difficulties.

Similar struggles were noted with respect to the cultural aspects of Tasha's mathematics lessons. She came into each post-teaching reflection session already thinking about what she would say to explain how the lesson had built on her students' cultural experiences. Each time she would share that she was not sure. For example, when reflecting on a lesson during her third year she said, "All last night all I could do was think about it . . . I don't know . . . I really tried to think . . . I don't know that culture was in it."

Tasha was able to explain that she felt in order to meet this goal she needed to find a way to connect to her students' out of school experiences. More than once she shared that one of her future goals was to make time to visit every child's home at the beginning of the school year. She felt making a positive connection with the family and talking to the child outside of the classroom might help her provide opportunities to build on each student's cultural capital in her mathematics lessons.

One of the reasons we suspect Tasha was willing to continue to take such big risks in her mathematics lessons is she seemed to realize how important it was for her students that her mathematics teaching change and continue to improve. There were many experiences from Tasha's past that brought this to her attention. For example, in her initial project interview she recalled an experience with a parent that really caught her off guard. Upon meeting the mother of a Black student Tasha said, "She asked did I know how to teach a Black boy. And it really—and that's always kind of stuck with me. . . . Because I always thought, you know, I didn't."

During DCY-2, Cohort 1 teachers read and discussed excerpts from the book *Radical Equations: Math Literacy and Civil Rights* (Moses and Cobb 2001). They also examined how the mathematics they taught in K-2 was foundational to algebra. This experience really hit home for Tasha. She shared in a presentation that it was the first time she realized her mathematics teaching was an important part of keeping the gate to algebra open for her students.

With Tasha we saw a woman who perceived the changes she had made to her mathematics teaching were very important. Even though she was well aware her lessons were not hitting the mark with her students, and she had not figured out how to capitalize on her students' culture within them, she kept at it. It was baffling that after three years in the project Tasha had not made bigger strides. However, her passion for the ideas espoused in NMD was obvious. Likewise, it was very clear she had no intention of giving up on them.

Relaying how a colleague had asked, incredulously, if she was still doing NMD, Tasha noted,

> I've told somebody I'm going to do it forever. They were like "you're going to do it forever?!" Then it was like, "I'm not going to go to Dreamkeepers forever, but it's always going to be a part of what I do." . . . I don't think they realized that it wasn't a workshop and it wasn't a training. It was a change.

Although we were unable to discern major changes in Tasha's mathematics teaching, by the end of the project we did see a teacher for whom it seemed certain she would continue to grapple with the important issues. She believed important strides had been made for the beginning of her journey.

KENDRA: THE CONSUMMATE PROFESSIONAL

Kendra was one of the most put together, neatly dressed, and well-kept teachers in the project. She always presented herself in a professional manner and it seemed the project's early focus on professionalism resonated most with her. Like Aisha and Tasha, Kendra felt she had experienced tremendous growth from participating in the project. However, certain elements, particularly her implementation of mathematics, were problematic.

As the youngest of six children in a religious family in an African American community, Kendra grew up in the South. Her parents were very involved in her schooling, often participating in school activities. Kendra had both African American and White teachers but felt more supported by the former because of their connection to the African American community in which she lived. She attended an ethnoracially diverse elementary school and majority White settings for middle and high schools. During her adolescence, Kendra met students from a variety of ethnicities, particularly students with Cambodian, Vietnamese, and Chinese heritage.

Because of her early experiences, Kendra felt comfortable discussing cultural diversity as well as interacting with others from different cultures. This helped to pique her interest in participating in NMD. She wanted to learn more about mathematics and how to combine it with culture. In addition, she wanted a better understanding of how her students learned and how she taught mathematics. She saw participating in NMD as resulting in a "paradigm shift" (a phrase she borrowed from a film about culturally responsive teaching shown in one of the first project retreats she attended) in that it changed her thinking about teaching and learning.

One area in which we saw substantial growth for Kendra was her understanding of culture and its impact in the mathematics classroom. Despite having grown up in a diverse environment, participating in NMD allowed her to understand that students' cultural backgrounds and prior experiences are important to the teaching-learning process. She stated, "If we [teachers] don't understand their culture and where they're coming from, and how they're taught, and how they understand math, it is hard for us to teach." Toward the end of the project, Kendra had rejected the colorblind perspective because it was not allowing her to truly see and understand the value in her students' cultural differences. She learned to acknowledge ethnoracial heritage as an asset. She explained:

> I often say, "I see color." Not just for my African American students. I see it for my Hispanic students, for my Asian students, for students of all cultures. I actually look forward to coming into my classroom and seeing color. I've actually taken that term out that "I don't see color." I do see color.

Although Kendra experienced growth in her understanding of culture and its impact in the classroom, she often vocalized that she experienced the most growth in the area of mathematics. She was proud of changes in her teaching that focused on incorporating students' prior knowledge. "Rather than formulating my own opinion on what they should know based on the way I'm teaching, [I] let the way they were taught shape and mold my teaching style." In addition, she even became more comfortable with mathematics personally. She expressed:

> Actually I've always had a thing for math. I've always loved math. I've always liked using different ways of showing how I understood math. I always tried with math . . . . I've always had just a love for math. My feelings not only escalated to the point where I really love math. And I don't mind—I would take all day teaching math if I could. I love math and I love watching the kid's eyes light up enjoying math as well.

Despite her self-proclaimed love for mathematics, Kendra's actual teaching of mathematics at times suggested that she did not have a strong conceptual understanding of certain mathematical concepts nor was she an effective teacher of mathematics. In the classroom, she often engaged students in simplistic mathematical tasks, offering few opportunities to challenge or extend their learning. For example, seven out of eleven of her observed mathematics lessons can be described as demonstrating low authenticity of mathematics (e.g., students finishing quickly, students focusing on tasks other than mathematics, etc.), and all eleven of her lessons featured low quality of teacher questioning of students (e.g., focusing on answers only, recalling facts, etc.).

In addition to being a weak mathematics teacher, Kendra also demonstrated that she was not a culturally relevant educator. She had a difficult time making connections between SBM and CRP. In her mathematics lessons, she rarely demonstrated any teaching practices that were in alignment with CRP. We identified only five instances, across eleven lessons, in which she integrated CRP subskills. The majority of her efforts to draw on culture were superficial at best (e.g., using ethnoracial-sounding names in word problems). Although, as indicated earlier, participating in the project helped Kendra to experience growth in understanding the impact of culture in the mathematics classroom, this growth was not sufficient to counter her inability to teach mathematics effectively. As a result, Kendra did not make substantial connections between mathematics and culture in her teaching.

Because of the difficulty she experienced in teaching mathematics, Kendra did not always easily relate to her students. Indeed, at times she presented a demeanor that appeared somewhat removed. For example, Kendra was required by her administration to write lesson objectives on the board. The presence or absence of the objectives was included as an element in teacher observations. As such, Kendra often spent considerable time in every mathematics lesson stressing the lesson's objectives and attempting to help the children understand the wording of the objectives rather than actually working on mathematics problems. In doing so, she presented the objectives in a mechanical fashion using language that was inaccessible to first graders. This gave the impression she was oblivious to the fact that she was not connecting with her students.

For her part, Kendra viewed use and explanation of the objectives as one of the most important elements of teaching mathematics. Surprisingly for us, she even saw this as a major takeaway from participating in the NMD project. When asked how her teaching mathematics had changed since participating in the NMD project, she explained:

> One of my major things is not always putting that objective on the board. Our administrators prefer us to list objectives. And they prefer it because they want the kids coming to the classroom knowing what they are going to be doing for that day. But I love for my students to tell me what they think we are going to do in math. That's something I want them to tell me. What they think we're going to do in math. And I look forward to them exposing to me what they know we're going to learn.

Although Kendra did not teach mathematics in the manner emphasized in the project, she was very receptive to the NMD emphasis on teachers as professionals. Often she raved about how NMD differed from other county professional development workshops. She stated, "I personally felt like I was treated like a professional all three years." She enjoyed attending project retreats and looked forward to learning from her NMD peers.

Through their participation in the project, she felt NMD teachers had a lot to offer their schools. Kendra wanted to share the knowledge she and the other teachers from her school had gained from participating in the project and suggested to her administration that the NMD teachers should lead some school-based staff development workshops. Kendra was really motivated by her experience in NMD and wanted to share what she had learned with others. After the NMD project ended, Kendra enrolled in a leadership academy project that was sponsored by a local university. It's likely NMD inspired Kendra to further develop her professional identity and her budding identity as a leader.

Overall, Kendra experienced growth in the area of culture but struggled in her teaching of mathematics, particularly the combining of culture and mathematics. Her participation in NMD helped her to develop an identity as a professional. She began to believe she was a professional mathematics educator. However, her implementation of mathematics in the classroom was problematic and indicated that in some ways she had developed a false identity.

HADLEY: CONSISTENTLY RESISTANT

Thus far, all of the secondary case teachers we have discussed viewed the NMD project positively and were effusive in their claims of having benefited from their participation. Although the majority of the teachers felt this way, there were a few who perceived certain components of the NMD project as problematic. In fact, we eventually learned that several of the teachers at Juniper found aspects of NMD to be quite troublesome. They frequently verbalized their discontent among themselves through mocking humor about an experience with a mathematics activity in which they participated at the first retreat. However, no teacher was quite as expansive in her commentary as Hadley. While she too was bothered by the first retreat's mathematics activity, she seemed particularly troubled by the culture focus of the project as was evidenced in her cultural autobiography and her end-of-year interview. It was her brash statements of discontent that helped her to be classified as a negative case.

Hadley was born in the late 1960s in the South to a conservative Christian family with five children. Her parents did not have a lot of money as indicated by her having to wear hand-me-downs and their shopping at thrift stores. Despite her family's financial difficulties, she did not recall going without the basic necessities. According to Hadley, her parents believed in working hard and they instilled in her a strong work ethic. She has been working since she was twelve years old.

Hadley grew up near a downtown area in a predominately White, low socioeconomic neighborhood. She attended a small, Christian school with White teachers because her parents did not like her predominately Black neighborhood school. Hadley explained, "My parents sacrificed to send us to a small church school to avoid having us be the minority White in our base schools." Hadley further explained that her parents were willing to do whatever it took for her and her siblings to attend the private school: "My mom didn't want to send us to the predominantly Black high school and middle school so she started cleaning houses for people to earn money."

Hadley had mostly negative perceptions of and interactions with people of color, particularly African Americans. She indicated that her neighborhood was located near "an impoverished Black community with a high crime rate" and mentioned that there were occasions where she felt threatened by "people of color" in her neighborhood. Hadley explained:

> Just a few blocks down the street was I remember huge groups of African Americans walking down our street on their way to the city pool during the summer. I thought they were very loud and unpredictable. On two different occasions, a group of Black men stopped their cars and tried to abduct my brother and I as we were playing by the street. I ran and ran feeling very afraid.

Although the majority of Hadley's interactions with people of color were negative, she did mention some positive relationships that occurred in her childhood. Hadley shared that a few elderly Black women attended her church and she "loved to sit with and talk to" them. In addition, she discussed how she enjoyed living next door to a Pilipino family and that they would often share their cooking with her family.

Hadley joined NMD because she wanted to expand the best practices in her mathematics teaching, which she described as "another tool in my belt." In addition, she was interested in pursuing a master's degree and thought participating in a project associated with a university would be a good way to learn about potential master's programs. Nevertheless, she was not receptive to differing perspectives on issues in which she felt knowledgeable. For instance, she questioned the project's implementation of mathematics inquiry activities at project retreats. She did not like being told that her answer was "wrong" or being asked to produce alternative explanations.[2]

Hadley especially questioned the project's interpretation and discussion of culture. At the time, Juniper had just finished a school-wide workshop on poverty in education and subsequent to this, as part of the project, the teachers were assigned to read a critique of the Ruby Payne poverty framework (Gorski 2008). This may have prompted Hadley to perceive the project focused too much on race instead of socioeconomic status. She explained:

> I guess there were times during the retreats that I just felt like in talking about culture it seemed like it was more towards just the skin color like White, Black Americans. I think I personally feel it's more social economic has a greater impact than just cultural. I think we need to respect each other but I don't know.

In addition to her discomforts over the project's focus on culture, Hadley disagreed with some of the specific topics that were covered in the retreats. When discussing racism in schools, she felt targeted as a White teacher and apparently often left retreats having felt "bullied." She explained:

> I felt like there were times during the retreats that I felt offended, that it's because I'm a White female teacher. Basically it's the White female teacher's fault that the African Americans are not up to par. You know that's—I don't know—that was offensive.

Drawing on this response from the final project interview, it is clear Hadley perceived White teachers were being blamed unfairly for the many problems in the education system, particularly those involving the achievement gap and African American students. Thus, Hadley felt a strong need to defend her position as a White teacher. Noting the dominance of White teachers in the profession, Hadley explained:

> To me that kind of borders on a dangerous edge because there is a shortage of teachers. And White female teachers are prevalent in the teaching field. And if White female teachers didn't rise to the occasion and teach we would be in a serious dilemma. You know I wish there were more African Americans teaching. And I wish there were more in the workforce.

Instead of focusing on issues of culture, particularly race, Hadley believed that teachers needed to concentrate on teaching their students to have a good work ethic and to demonstrate self-responsibility. Showing respect to all students regardless of their ethnoracial backgrounds and celebrating everyone's cultural contributions would create more inclusive classrooms. In fact, she perceived that focusing on culture (which she apparently defined as race exclusively) could even be detrimental in the classroom. She detailed:

> I mean there's no easy quick answer but it has to be where kids are motivated to learn no matter if they're Black, White, Hispanic. They have to be motivated to learn and to find ways to motivate them. It's not just giving them things but for them to feel pride. And I can understand how they can feel pride in a culture but I think sometimes if you focus too much on culture then they know they're—they know that their culture is maybe different or my culture is different. But if you focus too much on the culture then I have also seen that backfire like, well what are you going to do for me? Or have more of you do for me as opposed to I want to do my very best and I'm proud of who I am and I'm good at what I do. I think there's a fine line.

Curiously, despite her broad discomfort with the focus of the project (particularly the emphasis on culture in the teaching-learning process) it seems Hadley perceived NMD had been of value to her. She stated, "There are a lot of good things as far as just math talk and higher level thinking and getting kids to feel comfortable with math and providing multiple experiences and so forth." Hadley did admit that the cultural focus of the project was also valuable, but with a caveat. She explained, "The cultural piece is important also. I just think we're not going to get anywhere if you offend the teachers." Although Hadley was apparently dissatisfied with isolated features of NMD, it was perplexing that she felt the project was valuable, even the cultural aspect for which she offered a passionate critique.

JENNA: QUIET AND STILL WATER RUNNING DEEP

Jenna was a Cohort 3 first grade teacher from Juniper Elementary. In reflecting on her interactions in the retreats alone, we were challenged to say much about her. By her own admission Jenna was very shy. As a result she typically spoke quietly within small groups, and beyond that context she shared very little.

Jenna was not selected as a primary case because we had little insight into her thinking through her retreat interactions, lesson tallies, or lesson reflections. However, her participation in the project E-forums, written artifacts, and her final interview provided interesting and thought-provoking insight into the multiple ways Jenna was engaging in and thinking about the project content "behind the scenes," in a manner of speaking. It also gave insights into the ways in which she interacted with her colleagues.

Jenna describes herself as a White female from a small middle class farm town in the Northeast. She was in her third year of teaching when she joined NMD. Jenna was drawn to the project originally due to the mathematics focus, but it was the culture focus that really pulled her in and challenged her. A close look at Jenna's mathematics lessons reveals a teacher that was quite comfortable with the mathematics she was teaching. She had a relaxed discourse with her students, somewhat similar to Giselle. Jenna's students regularly posed thoughtful mathematical questions to her, as well as to each other. Her comfort with the mathematics focus of the project is evident in the coding of her lessons. Each lesson was rated as medium to high on both *authenticity of mathematics* and *overall quality of questions*. Her tallies—across all lessons—fell into two categories only, *Illuminating Thinking* and *Learning Connecting*. Jenna did not have any portion of any lesson coded using a CRP component of the lesson rubric.

Though there was little in her lessons that showed evidence of thinking about the role of culture in her teaching, we saw the beginning of ideas in Jenna's lesson reflections. She did not have what she felt were strong ideas about how she might have been able to draw on students' culture in the lessons we observed, but she was willing to engage in brief conversations about "possibilities" after the fact. For example, when her buddy triad was discussing another teacher's lesson and whether or not there was anything related to students' cultures that might have impacted their understanding in some way, Jenna was the first to quietly offer a suggestion. She noted, "I mean, language is a part of culture. And that might could be a problem." This suggestion she posed somewhat meekly, but among her buddy triad she was the only one who consistently offered thoughtful suggestions each time the question about integrating culture into mathematics lessons was posed.

Jenna's shyness was not only evident in her reflective interviews but during the project retreats as well. Barring times when she was specifically called upon to contribute, we were rarely sure what she was thinking during even our most challenging activities or discussions. Her quietness may have been due to her shyness, but it may also have been due to discomfort. In her cultural autobiography Jenna wrote,

> I am not comfortable talking about race and culture with someone of a different race than I. I hope someday I will reach that but right now I don't feel I have the right because I don't truly understand where they have come from and what they have been through. I feel I need to learn from them and listen to how they feel and what they have learned. . . . Almost like I don't feel the right to talk about it because I don't know what it is like.

As the project progressed, it became clear through the E-forum exchanges that Jenna was getting more comfortable discussing race and culture, at least in an asynchronous context.

Jenna participated in the eight online forum discussions significantly more than all but one of the Cohort 3 teachers. An analysis of her forum contributions revealed she wrote about race and culture in nine of her twenty-five posts. Most of those happened during the second half of the year. For example, in the February forum, in response to a colleague who had complained about a retreat activity that was not as explicitly focused on culture in specific math lessons, Jenna noted, "I think the reflection of culture on mathematics would be hard to replicate on a large scale because its [*sic*] very difficult to see in classrooms that are not diverse and hard to believe that many of the things are still happening in today's society."

It is possible this particular medium (i.e., online asynchronous exchanges) might have made Jenna more comfortable participating in discussion about culture issues that she perceived as difficult. Ironically, it was through this forum where we first saw a hint of her frustration with her Juniper colleagues. Like the forum post above, there were other posts that indicated Jenna's discomfort over the lack of seriousness with which at least some of her colleagues were attending to the cultural focus of the project. When asked at the end of the project what was most disappointing, she explained,

> I guess some of my biggest disappointments have been the way that the other teachers looked at the culture like there's nothing wrong. I don't know. I hate to say that but I feel like that was a little bit more disappointing. Like why aren't you being open minded and kind of seeing what she's saying . . . . I guess I just didn't really like that.

In her final interview, Jenna reflected on what she learned through participating in the NMD project. She shared that she appreciated being reminded of best practices for mathematics instruction and found it helpful to discuss them and reflect on them on a regular basis. However, she notes that by the end "I started taking more of the cultural aspect from it than the math." While her growing appreciation for the importance of attending to the facets of CRP had not played out in her classroom during the span of the project, Jenna's case is different from the expected pattern among the NMD participants. She was so quiet and unassuming that we had little sense of how she was responding to the intervention.

Even so, we found out later that this very shy young woman began to stand up and even nudged her colleagues to engage seriously with issues surrounding the impacts of culture in the teaching-learning process. Jenna's growing willingness to discuss race and culture, as well as to challenge her colleagues through the E-forum, indicates that she has at least begun to take on characteristics that are consistent with a culturally relevant mathematics teacher identity.

CONCLUSION

With a few notable exceptions, the majority of the teachers seemed to greatly appreciate the experience that was the NMD project. They reported having grown in their understanding of the importance of tending to culture as part of their own pedagogies even though there was scant evidence of its integration in their actual mathematics lessons. Likewise, while the teachers appreciated the differences between procedural and conceptual mathematics understandings, how to make those distinctions in the context of their own pedagogy was a significant challenge for most. In short, analysis of our data revealed there was a definite theory/practice chasm for at least some of the teachers when it came to merging mathematics and culture in primary grade teaching and promoting deep understanding for young students.

The stories of the primary case teachers left us with a sense of affirmation about the overall success of the NMD project, whereas those of the secondary case teachers challenged us to reflect on the outcomes and recognize that in various ways the experience was confounding for some participants. Most of the secondary case teachers showed scant evidence of drawing on culture in the context of mathematics lessons. This may suggest that they did not fully embrace a basic premise of the study. We explore this issue in the final chapter.

NOTES

1. Various writers have reported that for many preservice teacher candidates as well as in-service teachers, translating knowledge about cultural diversity into transformative classroom practice presents particular challenges (Brown 2004; Hodge and Cobb 2010; Sleeter 2008). We therefore expected some disconnection between the clear understandings Aisha articulated during project retreats and what occurred at her school site. But we did not expect the *extent* of disconnection we found upon analyzing Aisha's post-teaching reflections and interviews.

2. At the end of the project we learned Hadley, and several of the teachers from Juniper, reacted negatively to the Candy Factory mathematics simulation activity from the first project retreat. As part of the critical thinking reflection part of the activity, teachers were required to present solutions. After Hadley's group presented, co-PI McCulloch suggested that they "dig deeper" to find alternative explanations. Unfortunately, this response was interpreted as a rebuke of the group's solution. Comments shared during the end-of-year interview made clear some teachers had taken offense at co-PI McCulloch's response. According to Hadley, a running joke among the Juniper teachers throughout the year was that their solution at the first retreat had been called out as "wrong."

Chapter 6

Toward a Liberating Professional Identity

This final chapter is our reflective analysis and critique of the NMD project organized around an adaptation of the questions used in our post-teaching reflection sessions. Recall from chapter 2 that the interview protocol for those sessions included a series of prompts, used in a mainly linear fashion by the RAs, to facilitate the teachers' reflections on the intentions for and outcomes of their mathematics lessons. The discussion in this chapter diverts somewhat from that structure as we recognize the prompts were not mutually exclusive, but rather in some cases they overlapped. Our hope is that through the use of this structure we will provide novice researchers a useful glimpse at the multilevel reflection necessary at the conclusion of a longitudinal study.

We begin, therefore, by revisiting the project objective and explaining what it would have meant for the teachers to integrate components of multiculturalism (CRP) into their mathematics instruction. Next, drawing on our data sources, we describe and analyze the overall gains we attribute to the teachers' participation in the project. Then, we critique manners in which focused commitment to implementation of our research plan inadvertently hindered the full realization of some aspects of the study. Herein we incorporate lessons we learned that could be used by others in the conduct of future longitudinal, school-based studies. Finally, we draw on the project's mathematics/multiculturalism dual focus to theorize affirmative intersectionality in the broad structure of professional development studies like NMD that target the goal of reflective, ethical teaching practice in contemporary culturally diverse schools.

MULTICULTURAL SENSIBILITIES IN THE TEACHING OF PRIMARY GRADE MATHEMATICS

Each post-teaching reflection session began with a two-part question: *What mathematics ideas were you trying to teach, and what does it mean for someone to understand those ideas?* It seems fitting, therefore, to start our analysis and critique of the NMD project with a variation on that query.

What Ideas Were Teachers to Understand About the Mathematics/Multiculturalism Connection?

There were three central ideas for the teachers to understand as a result of their participation in NMD. The first was SBM teaching is superior to traditional instruction in regard to promoting deep mathematical understanding. Second, teaching grounded on the ways of being and the cultural realities of ethnoracial minority students enhances their interest, engagement, and academic achievement. Independent studies in mathematics education (Ball, Lubienski, and Mewborn 2001; Hiebert and Carpenter 1992; Kazemi and Stipek 2001) and in multicultural teacher education (Gay 2002; Ladson-Billings 1994, 2006; Lee 2001; Lee, Spencer, and Harpalani 2003) bolster the two forgoing ideas.

In recent years, mathematics education and multicultural education have been combined to refine understanding of the complexities of mathematics teaching and learning in contemporary culturally diverse schools. Thus, a third idea our teachers were to understand was integrating SBM techniques with a strategy for cultural relevance in teaching can enhance deep understanding of mathematics for all students, and particularly among those from ethnoracial minority backgrounds (Enyedy and Mukhopadhyay 2007; Gutstein et al. 1997).

Understanding these three ideas meant the teachers were to exhibit certain capacities (Knight and Wiseman 2009). For NMD, the notion of capacity extended beyond discrete instructional techniques, although these too were central in our assessments of the teachers' understanding of the three central ideas. Still and all, we operationalized the concept of capacity to include teachers' depth of understanding of mathematics content knowledge. Also, we included interest (as articulated in written and oral reflections on instructional practices) in utilizing children's cultural ways of being to enhance their mathematics understanding.

Shulman (1998) noted, "Teachers must be educated and socialized to develop dispositions toward inquiry, reflection . . . , and an orientation to direct their attention at the underlying intellectual and motivational processes of the child" (p. 514). The disposition NMD advanced was one grounded in commitment to equity and opportunities to learn for all children. Its ultimate outcome was focused on high academic achievement. We wanted teachers to want to know about (and make use of) the out-of-school experiences with mathematics their students encounter regularly. Doing so, we reasoned, would enhance academic achievement.

Teachers were expected to exhibit positive dispositions toward teaching as complex intellectual activity as well as sensitivity to the reverberating impact of high quality early mathematics instruction. In a phrase, the NMD project promoted a certain *professional identity* among the teachers. It was one aligned with broad tenets of CRP (Ladson-Billings 1994, 2006).

What Was the Integration of SBM and CRP Supposed to Look Like in Lessons?

Teachers want to know what cultural relevance looks like; however, in a response to feedback over the years, Ladson-Billings (2006) proffered "'doing' is less important than 'being'" (p. 41). It is likely this response strikes some as enigmatic, but to us it means the substance of CRP is found in the broad stance a teacher assumes in service to assuring academic success for all her students. Further, we take this to mean that the stance a teacher takes is more critical than the particular set of instructional tactics she may choose to execute in any given instructional moment.

Still, we felt intense pressure from the NMD teachers to specify, if not prescribe, what we meant by cultural relevance as part of SBM teaching. This pressure to operationalize ideas became even more acute, and understandably so, when we assigned RAs to code videotaped mathematics lessons for the presence of both CRP and SBM (DeCuir-Gunby, Marshall, and McCulloch 2011). Therefore, drawing broadly on previous scholarship (Boykin et al. 2005; Boykin et al. 1997; Howard 2001, 2003; Ladson-Billing 1994; Lee 2001; Lee, Spencer, and Harpalani 2003; Moemeka 1998; Tucker and Herman 2002) we identified the presence of CRP in a lesson when teachers exhibited evidence of certain discrete teaching acts. These acts we operationalized as *Language Connecting*, *Relevance Making*, *Cultural Connecting*, and *Communalizing*.

For example, *Language Connecting* was identified when a teacher used children's own nonmathematical language in the conduct of a mathematics lesson with the purpose of affirming a child's own ways of articulating understandings of mathematical ideas. On the other hand, *Relevance Making* occurred when the teacher helped a child make or recognize associations between a mathematics task or idea introduced in school, and how knowing that idea will be helpful in school *as well as* nonschool situations.

Cultural Connecting was defined as making direct reference to a child's home/out-of-school experiences to affirm and illuminate linkages to particular school mathematics content. Finally, *Communalizing* occurred when the teacher incorporated different cultural ways into the learner socialization mores of the class. This was exemplified by teachers' creating a classroom ethos wherein students recognize they are responsible for and obligated to the success of each other.

We defined SBM discrete skills in terms of *Learning Connecting* wherein teachers identified explicit associations and relationships between and among ideas being taught in lessons. *Illuminating Thinking* was present when a teacher gave directives or posed questions intended to highlight a child's understanding. Also, it was recorded when a teacher's verbal directive was used to require a child to more fully articulate ideas behind responses and answers given during oral exchanges. *Affirming Multiple Representations* was present when teachers accepted and encouraged a child to represent mathematical ideas in various ways; whereas *Extension of Tasks* occurred when teachers expanded upon a task that initially did not present sufficient challenge or cognitive loading.

In the long run, our decision to identify and operationalize discrete teaching acts consistent with merging CRP and SBM proved quite useful. At the same time, however, we recognized that the whole of the integration we sought between these two foci could never be captured fully by the sum of the individual teaching parts identified in any given mathematics lesson.

WHAT UNDERSTANDINGS DID THE TEACHERS ACTUALLY ACQUIRE? HOW DO WE KNOW?

As is clear from our primary and secondary cases, the questions we pose at this point are very complex ones to consider. Therefore we begin with what teachers learned about mathematics and the teachers' dispositions about merging CRP/SBM. Next, we identify the ways in which the teachers demonstrated their knowledge of CRP/SBM in their mathematics teaching.

Our challenge in exploring that issue, however, is that teachers like Jenna and Tasha illustrate that what we saw in some lessons did not provide a full picture of the depth to which some teachers actually came to understand CRP or SBM. Finally, we anticipated that changes in instruction would result in an increased understanding of mathematics for the project teachers' students as well. Therefore, we also present results of student outcomes.

Teachers' Knowledge of K-2 Mathematics

By the close of the final data collection year we observed several changes among the teachers. For example, many exhibited a somewhat keener understanding of the culture-teaching-learning relationship. Others demonstrated gains in their ability to receive and make use of feedback obtained through peer observation and post-teaching reflections. Particularly noteworthy was the change in the teachers' mathematics content knowledge.

A question we posed for the study was *how do K-2 teachers re-organize their teaching . . . in such a manner that they demonstrate the ability to decipher the intended mathematics of standard K-2 mathematics curriculum?* This question spoke to the teachers' own mathematics content knowledge. At the same time, however, teacher mathematics knowledge is quite complex and difficult to measure. Therefore, to find answers we analyzed data collected using our Teacher Mathematics Questionnaire (TMQ) (see Appendix C).

All items on the TMQ align with the mathematics strands as defined by the NCTM (2000) including number and operations, algebra, geometry, measurement, and data analysis; hence, the development of the TMQ focused on the mathematics normally taught in K-2 classes. In addition to typical mathematics problems, the TMQ included items in which teachers analyzed student work samples. Likewise, items were structured for teachers to demonstrate ability to recognize common misconceptions, and select age-appropriate helpful representations of the mathematical ideas.

Our goal in structuring the TMQ around these ideas was to discern the degree to which the teachers themselves understood the mathematics they are required to teach, as well as their appreciation for the complexities of student thinking. Therefore, each item required teachers to provide a solution *and* an explanation for that solution. The assessment rubric we developed for the TMQ measured both correctness and quality of explanation with the latter being scored along a continuum that included high, medium, and low evaluation rankings.

In order to analyze the pretest and posttest data, we used both parametric and nonparametric statistical procedures as appropriate for the size of the three cohorts of teachers. The population of all three NMD cohorts (as well as Cohort 3 by itself) was large enough in size to use parametric measures (i.e., t-tests). However, since Cohort 1, Cohort 2, and the Control Group were smaller in size ($n = 8$, 10, and 16, respectively), a Wilcoxon signed rank test was used. The tests revealed all three cohorts showed significant improvement on the TMQ (see Table 6.1).

Further, all groups showed significant increase in the center score (i.e., mean for paired t-test versus median for Wilcoxon) on the TMQ.

It is important to note that the Control Group had a much higher center score on the pretest than the other groups. This center score increased only by slightly more than six points on the posttest. The significant increase in scores is accounted for by the tightening of the scores around this center, not by the movement of the center itself. In contrast, the NMD teachers' center score

Table 6.1 Comparison of Pre- and Posttest Centers on Teacher Mathematics Questionnaire

	Center/Spread	*Significance*
NMD (n = 49)		
Pretest	M = 52.61, SD = 18.39	Paired samples t-test
Posttest	M = 71.32, SD = 17.65	$t(48) = -7.352$, $p < 0.001$
Cohort 1 (n = 8)		
Pretest	Mdn = 36.27, R = 39.22	Wilcoxon signed rank test
Posttest	Mdn = 81.37, R = 45.10	$W = 36$, $p = 0.007$
Cohort 2 (n = 10)		
Pretest	Mdn = 46.08, R = 49.02	Wilcoxon signed rank test
Posttest	Mdn = 66.67, R = 62.75	$W = 53$, $p = 0.003$
Cohort 3 (n = 31)		
Pretest	M = 58.32, SD = 16.88	Paired samples t-test
Posttest	M = 71.09, SD = 17.99	$t(30) = -4.37$, $p < 0.001$
Control Group (n = 16)		
Pretest	Mdn = 62.17, R = 43.58	Wilcoxon signed rank test
Posttest	Mdn = 69.87, R = 29.48	$W = 122$, $p = 0.003$

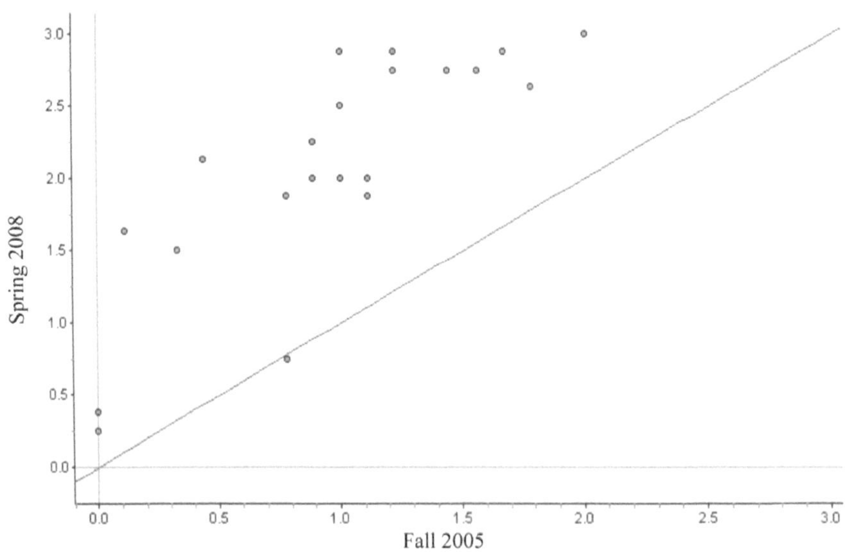

Figure 6.1 Comparison of Mean Quality of Explanation Scores for Cohort 1: Fall 2005–Spring 2008

increased by almost twenty points while also tightening to within eighteen points of the center on the posttest.

For each of the cohorts we also looked at the mean quality of explanation scores from the first to the last administration of the TMQ at the item level. Findings appear in Figures 6.1, 6.2, and 6.3. Each point on the graphs

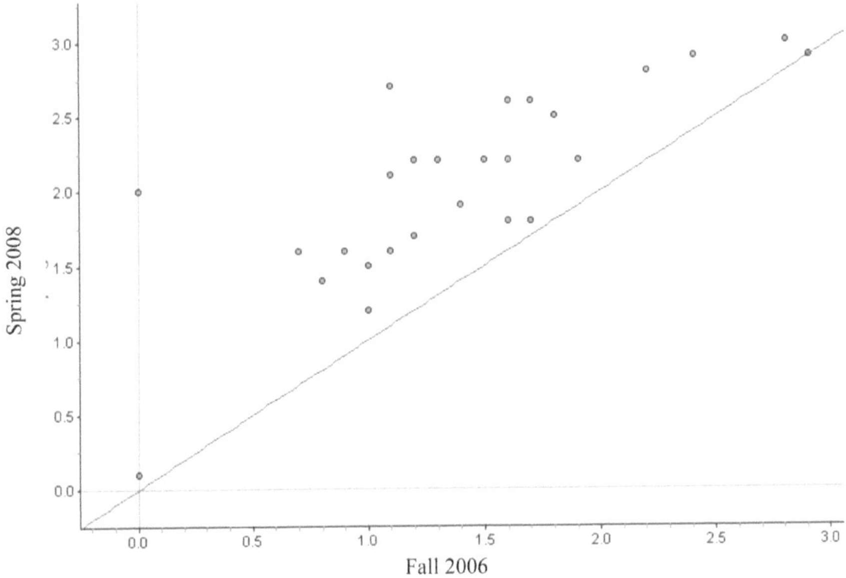

Figure 6.2 Comparison of Mean Quality of Explanation Score for Cohort 2: Fall 2006–Spring 2008

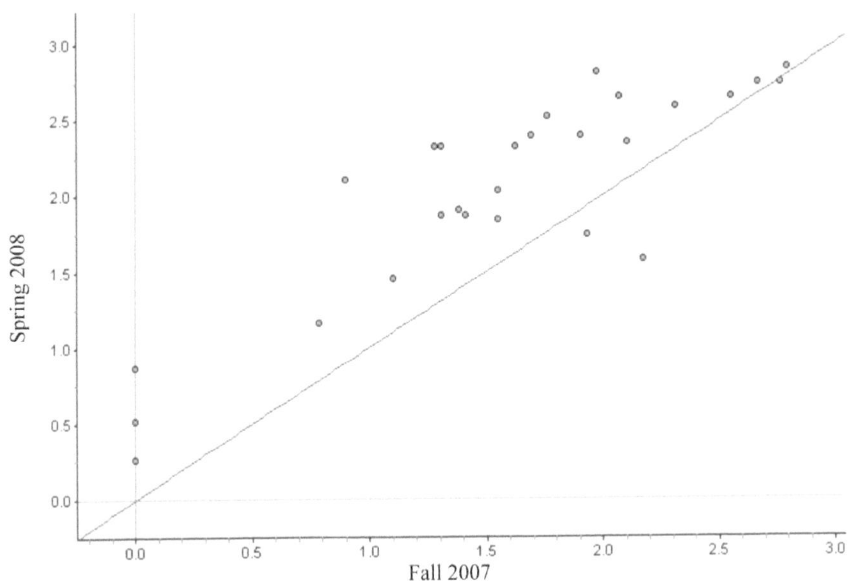

Figure 6.3 Comparison of Mean Quality of Explanation Score for Cohort 3: Fall 2007–Spring 2008

represents a particular item on the TMQ and the lines signify where the point would fall if there had been *no change* between the first and last administrations. Items in which an increase in the mean quality of explanation occurred for a cohort overall appear above the line, whereas those for which the mean quality of explanation decreased occur below the line.

As shown in Figure 6.1, from the first to the last administration of the TMQ the quality of explanation for Cohort 1 improved on all items. The results for Cohort 2 (Figure 6.2) show that the teachers either maintained or improved their quality of explanation on all but two items. Similarly, the results for Cohort 3 (Figure 6.3) show that the teachers either maintained or improved their quality of explanation on all but three items.

Based on the TMQ results, the NMD teachers grew with respect to their knowledge of the big ideas in K-2 mathematics. (See Appendix F for additional discussion of our analysis of findings from the TMQ.)

Teachers' Dispositions Toward Integrating CRP and SBM

All project teachers completed the *Teacher Dispositions Survey* (TDS), an instrument developed for NMD to assess the nature of change in the teachers' dispositions or orientations toward integrating multiculturalism/CRP in mathematics teaching and learning. (See Appendix D for a complete description of the TDS.) Using the DCY-3 (posttest) data, a series of ANOVAs were conducted, exploring the *group means* by treatment group (NMD Cohort teachers or Control Group teachers) for each subscale. A significant difference ($F = 2.964$; $p = 0.039$) was found between the means of Cohort 2 and the Control Group on the *Affirm/Value Community* subscale. The mean for Cohort 2 was 3.81, while the mean for the Control Group was 3.31.

This finding suggests that Cohort 2 teachers demonstrated more positive dispositions toward the nature and impact of children's outside-of-school communities, and by extension were more sensitive to issues concerning mathematics learning and the impact of culture thereon than the Control Group teachers. It also supports one of the NMD hypotheses that Dreamkeepers (in this case, Cohort 2 teachers) would develop knowing-in-practice that is more consistent with the project intervention than Control Group/standard practice teachers.

In addition, a significant difference was found between the means of the NMD teachers as a whole and the Control Group on both the *Parental Role in Schooling* ($F = 4.19$; $p = 0.045$) and *Affirm/Value Community* subscales ($F = 5.23$; $p = 0.026$). Means for the NMD teachers are 4.39 and 3.62 on the *Parental Role in Schooling* and *Affirm/Value Community* subscales, respectively, whereas means for the Control Group are 4.03 and 3.31 on the *Parental Role in Schooling* and *Affirm/Value Community* subscales, respectively.

This again suggests that at the end of the project, NMD teachers demonstrated more positive dispositions toward parental role in schooling and the notion of affirming and valuing communities from which children come. By extension this finding would suggest NMD teachers were more sensitive to issues concerning mathematics learning and the impact of culture thereon than the Control Group teachers. Likewise, it supports the notion that the NMD teachers exhibited some development of knowing-in-practice that was more consistent with the project's focus than the Control Group teachers, suggesting that the NMD research intervention was effective.

Facets of CRP and SBM in Action During Mathematics Lessons

Based on our analysis of their mathematics lessons and lesson reflections, we found the NMD teachers incorporated some aspects of CRP and SBM into their pedagogical practices. These practices included aspects of connecting children's out-of-school experiences to school mathematics, promoting high expectations for all, and valuing the children's mathematical thinking. It was through identification of such actions that we are able to make claims about what the teachers learned that directly affected their mathematics teaching.

The actions most tightly aligned with connecting to students' out of school experiences were coded as *Cultural Connecting*, *Language Matching*, and *Relevance Making*. In the 482 coded lessons, 149 instances of such connections were identified. These codes were spread over all three cohorts and forty-two teachers. We found that 92 percent of the codes appeared in lessons that were taught in the latter part of each teacher's experience in the project. This suggests that for these teachers, the longer they were in the project (and the more time they had for exploration), the more frequent instances of CRP appeared in their lessons (McCulloch and Marshall 2011).

In addition to the explicit instances of CRP identified through use of our lesson analysis rubric, an analysis of the reflective sessions revealed most of the teachers also considered their students' out-of-school experiences in the *planning* of their lessons. Enacted lessons reflected decisions teachers made regarding how to best structure the lesson as a whole with the students' experiences in mind.

When reflecting on a task some teachers purposefully chose the context of problems to match children's out-of-school experiences. For example, a Cohort 3 teacher chose to include buttons for a task, noting, "Some of the students could relate. You know, they had said, 'Oh, my grandmother has a box of buttons.' And they were really interested in the buttons." While this teacher was relatively certain most of the children were familiar with buttons, her assumption about the extent of their familiarity was obviously measured. She went on to explain:

> We talked about how buttons are sewn on. And we even looked at how people used buttons—you know, in the past. We had some pictures from a book that we had read. So [I'm] just kind of building on their background experience.

The most common code identified on the mathematics lessons was *Illuminating Thinking*. From a SBM perspective, we found this to be very promising. In 2014, NCTM released the document "Principles to Actions: Ensuring Mathematical Success for All." One purpose of this document was to help teachers identify important instructional practices (i.e., those consistent with SBM) that help to ensure learning for all students. The chapter "Effective Teaching and Learning" identifies facilitating meaningful mathematical discourse and posing purposeful questions as key teacher actions. These actions were captured by our *Illuminating Thinking* code.

We identified over 2,200 *Illuminating Thinking* instances across all cohorts. There was a significant difference between the project teachers ($M = 4.17$, SD = 4.43) and the control teachers ($M = 2.55$, SD = 2.74), $t(424) = -2.772$, $p = 0.006$, clearly revealing the former integrated purposeful questioning into their lessons more frequently. Even so, as we saw with the case of Giselle, our coding did not pick up on those teachers who had developed more sophisticated classroom discourse practices (e.g., students asking each other questions of this type).

Impact on Students' Mathematics Knowledge and Strategies

The NMD teachers touched the lives of over 2,300 students during the span of the project. Based on the Student Mathematics Surveys (Appendix E), we found children in classrooms of NMD teachers grew in their mathematical understanding. Our findings here are important to teacher education, as there are very few examples of studies that trace the impact professional development projects have on student achievement (Yoon et al. 2007).

Using a pre/posttest design, the first and second grade students of all Cohort 3 teachers and the Control Group teachers were assessed. (Recall from chapter 2 that the Control Group was not formulated until DCY-3.) A paired *t*-test was used to compare the means on the first and second grade questionnaires. As expected both Cohort 3 students and the control group students grew in their mathematics understanding from the beginning to the end of the academic year.

Comparing across groups we found a significant difference between pretest scores, but not between posttest scores for both the first and second grades (Table 6.2). The project first graders had a significantly lower pretest mean (0.5059) than the control group (0.6048), but by posttest there was no significant difference between the means. Thus, we conclude that the students

Table 6.2 Across Group Mean Differences on Student Questionnaires

Group	*Mean*	*SD*	*DF*	*t-Value*	*Pr > |t|*
First Grade pretest			221	3.23	0.0014
Cohort 3	0.51	0.24			
Control Group	0.60	0.22			
First Grade posttest			221	−0.23	0.8216
Cohort 3	0.73	0.18			
Control Group	0.73	0.21			
Second Grade pretest			194	4.08	0.0001
Cohort 3	0.38	0.19			
Control Group	0.53	0.19			
Second Grade posttest			194	1.90	0.0586
Cohort 3	0.54	0.22			
Control Group	0.61	0.20			

in classrooms of project teachers showed more growth on the questionnaire than did the control group for first grade. The same was true for the second grade students. Project second graders had a significantly lower pretest mean than the control group (0.3776 and 0.5272, respectively) and showed no significant difference by posttest.

In the end, we are confident the teachers grew in their own knowledge of K-2 mathematics for teaching. Likewise, they grew in the ways they enacted practices consistent with CRP and SBM. Lastly, their students grew in both their mathematical knowledge and sophistication of strategy usage.

WHAT DID WE DO THAT HINDERED OUR RESEARCH GOAL?

When reflecting upon our project, we thought about the various obstacles we encountered throughout the data collection process. As in any research study, we encountered a few complications and are obligated to discuss their potential impact on our findings. Our discussion of the primary case teachers in chapter 4 illustrated how several teachers exhibited varied degrees of success in integrating CRP and SBM. Contrarily, as described in chapter 5, many of our secondary case teachers experienced difficulty in integrating CRP and SBM.

A portion of these teachers' lack of success can be attributed to some of the methodological choices made throughout the project. Reflecting upon factors that may have hindered our research process, we focus specifically on the quality of the instruments used, the thoroughness of the RA and data collector trainings, and the timeliness of the data collection and analysis procedures.

The Quality of the Project Instruments

As described throughout the earlier chapters, we used a variety of quantitative and qualitative data sources, many of which were created specifically for the project (e.g., lesson analysis rubric, interview protocols, reflection interview protocols, etc.). One issue we encountered is that in creating our own project instruments, there was often insufficient time and inadequate opportunity to pilot them prior to the start of the study. Although this was the case, the creation of all instruments was grounded in the research literature.

In addition, several of the instruments, particularly the interview protocols, were meant to be emergent and structured around what was happening in our study. Great care was taken to ensure all data was carefully analyzed as indicated by our creation of codebooks for our interviews (DeCuir-Gunby, Marshall, and McCulloch 2011) and videotaped mathematics lesson analysis rubric (DeCuir-Gunby, Marshall, and McCulloch 2012).

Despite the cautions in creating instruments and analyzing data, we recognize that at times the instruments themselves failed to reveal the full breadth of the data for which they were designed and intended to capture. For example, to determine how teachers integrated aspects of SBM and CRP in lessons, we assessed classroom practices. This involved creating the lesson analysis rubric to capture subskills of both SBM (e.g., *Illuminating Thinking*, *Learning Connecting*, etc.) and CRP (e.g., *Language Connecting*, *Relevance Making*, etc.) that may have been incorporated in the teachers' communications during their lessons.

After analyzing the videotaped mathematics lessons, however, we discovered the rubric was not sufficiently sensitive to capture all relevant behaviors in the classroom. The situation with Giselle is illustrative. Recall there were few tallies for *Illuminating Thinking* in her lessons, an important subskill of SBM.

Not until later did we identify a plausible explanation. Giselle often engaged in CGI teaching practices, which focused more on student interactions. As such her teaching often featured students working in small groups and asking critical questions among each other. Since our instrument was not designed to focus on interactions between and among students, we missed this relevant information.

Although we attempted to create project instruments that would facilitate the collection of thorough and detailed information, there were instances when use of those instruments prevented us from capturing the most necessary information. We used the same reflection session interview questions after every interview. In hindsight, we recognize that we should have changed the reflection session interview questions frequently in order for teachers to focus on different issues within their teaching. Kendra, for instance, should have been asked to reflect upon her focus on the use of the lesson objectives, and to critique its impact on student learning.

Also, we probably should have used a different approach in creating the interview questions themselves. Specifically, we should have scaffolded the interview questions to make them more personally relevant to particular teachers. For a teacher like Aisha who was very aware of CRP, yet not incorporating it into her teaching, different questions may have alerted her to the inconsistencies between her knowledge of CRP and her actual teaching practices.

We do understand the need to use the same questions across participants in order to maintain consistency. On the other hand, as researchers, we should have been more receptive to the dynamism of the research process. In the context of a professional development project, in hindsight, we realized the questions should have evolved. When growth is promoted, there is an expectation that participants will demonstrate change in their thinking. Adapting the questions to the changing contexts could have helped to further promote the teachers' overall development.

For the final interview, we did create alternate questions for the different cohorts as we wanted to assess teachers' varied experiences in the project. However, it would have been helpful if we had taken additional effort to individualize the interview questions. For instance, Hadley was resistant to aspects of the multiculturalism/CRP focus of the project, but her responses to interview questions were opaque to the point that we could not pinpoint specific activities that had caused her concern.

More carefully worded questions may have teased this out. In other words, in the end we recognized that tailoring interview questions to the particular issues that emerged for individual teachers may have yielded more relevant information about each teacher.

Training the RAs and Data Collectors

As noted in chapter 2, a total of twelve RAs and two data collectors served on the project in a variety of capacities. They came to the project with various backgrounds and experiences (e.g., teaching experience, research experience, etc.) that impacted their ability to collect project data thoroughly.

For instance, one data collector was a retired veteran teacher with substantial experience observing teachers in the classroom. When conducting post-teaching reflection interviews, she asked relevant questions that focused on occurrences in the teachers' mathematics lessons. Yet another RA was a doctoral student in psychology with no K-12 teaching experience. Post-teaching reflection sessions conducted by this RA demonstrated far less detail due to her lack of familiarity with teaching. This uneven range of experiences resulted in different levels of completeness in the data collected for each teacher.

We provided training on how to utilize project instruments through the implementation of standardized training protocols. This included in-depth training on how to conduct classroom observations, write field notes, code interviews, score the lesson analysis rubric, and transcribe teacher communications using the rubric. Despite these efforts, there were inconsistencies in how the RAs and data collectors engaged in their collection duties. And this lack of consistency inadvertently impacted the experiences of some of the teachers in the project.

For instance, it occurs to us that had Aisha been asked about the discrepancies between her multiculturalism/CRP knowledge and her implementation of CRP in the classroom, she may have worked to bring the two into closer alignment (see Marshall, DeCuir-Gunby, and McCulloch 2012). We recognize the inconsistency in the data collection limits the claims we can make about the experiences of specific teachers.

One area in which there was considerable variation was in the creation of field notes. For some teachers there were detailed field notes while for others the field notes were sparse. Devoting more time to practicing the writing of field notes as well as examining how they *should* look would have been helpful for the RAs and beneficial for the project. In fact, we now realize that providing refresher discussions for the research team members on all the different collection techniques used in the project would have been helpful.

Another area where variations occurred was with the conduct of interviews. Some RAs attempted to probe for deeper responses while others simply followed the interview protocols with little or no follow-up questions. While we created a detailed protocol for the RAs to follow, they were encouraged and trained to ask follow-up questions to probe deeper when teachers provided interesting or incomplete answers.

For example, in the closing interview, Hadley made several inflammatory statements regarding culture. Her comments seemed to reflect considerable misunderstanding or misinterpretation of the concept; however, few follow-up attempts were made to allow her to clarify statements. It is possible that if Hadley had been pressed to clarify her statements, important insights might have been gained. Moreover, her insights might have been useful in the structure of some retreat activities.

In addition, consistent probing could have been used to help teachers to reflect more critically on their teaching practices. Isabel did not talk much about the impact of culture in her closing interview despite the fact that she had implemented a variety of culturally relevant practices in her teaching.

Data Collection and Analysis Timelines

Our original plan for the project was to use an emergent research design where we engaged in continuous data analysis to inform the various components of the project. This approach would have required us to analyze data soon after it had been collected. However, because of many unforeseen circumstances (see chapter 2) we were not able to implement the original plan. Instead, we implemented a more traditional approach where analysis was delayed until all data had been collected. Doing so significantly impacted the structure and implementation of the project retreats as well as interpretations we have made from the data.

Using the original emergent approach to data analysis would have allowed us to shape the project retreats around the experiences of our project teachers. The retreats would have reflected a degree of immediacy and more direct relevance to issues the teachers were encountering in their classrooms or schools. We learned from the E-forum discussions and interviews, for example, that Jenna perceived the teachers at her school (Juniper) were not receptive to issues of culture. Hadley's interview corroborated this point. Had we been aware of this information sooner, it could have been used to guide retreats and possibly create more fruitful conversations and activities surrounding multiculturalism/CRP and its connection to mathematics teaching.

Similarly, we could have used the data to help challenge the teachers' classroom practices. Specifically, early analyses of the classroom videos as well as the reflection interviews may have provided useful insights into what was occurring in the teachers' classrooms. In the cases of Aisha, Tasha, and Kendra, for example, we could have helped them to identify inconsistencies and/or work to address challenges in their teaching.

Each of these teachers was receptive to making changes in her pedagogy, and would have welcomed assistance to help with more constructive analysis and improvement of her lessons. Likewise, early analysis of the data could have been used to challenge our most successful teachers and further promote their professional development. Specifically, we needed the video evidence to help show Johnetta and Arlene that they were successfully integrating CRP and SBM and to help them determine ways in which they could continue to grow in the classroom.

In summary, we have taken this somewhat unusual turn in this part of our reflections on NMD (i.e., detailing missteps) to demonstrate that after a study has been completed, it is critical for researchers to examine how particular methodological choices may have inadvertently hindered aspects of the research process. At the same time, however, we want to underscore that our data collection and analysis procedures were greatly affected by time and personnel constraints.

We made interpretations from a variety of quantitative and qualitative data sources that were created specifically for our project. In hindsight, we recognize some of our choices may have influenced our findings in manners that may have been different if alternate choices had been made. Ultimately, we approached our data collection and analysis with integrity and fidelity, realizing that many of the problems we encountered were indicative of school-based research in general, and longitudinal research specifically.

REFLECTIONS ON A DREAM

Based on our reflective analysis and critique, we conclude that in many respects we met the objectives for the NMD project. Teachers increased their mathematics content knowledge and they incorporated features of CRP and SBM into mathematics lessons on a fairly regular basis. Moreover, in the final year of the project when we collected student data, our findings show that students in the classrooms of NMD teachers[1] performed better on our mathematics assessments than did students in the classrooms of control group teachers. Nevertheless, we finished the project with the realization that the essence of the mathematics/multiculturalism merger we sought had remained elusive for all but a few of the teachers. The majority stumbled time and again over the broad question, *How did you draw on and make use of students' cultures and extant experiences to promote understanding of the mathematics ideas you were trying to teach?*

Through the benefit of hindsight we believe the teachers struggled with this question for two interrelated reasons. On the one hand, many never fully accepted the premise that mathematics is culturally imbued. In this regard, similar to Sleeter's (1997) findings, it seems "as long as they viewed mathematics as a decontextualized and sequenced set of skills . . . [they] were not [inclined to be] asking questions about such matters as how children use numeracy in their daily lives and how diverse peoples historically created mathematics" (p. 691).

By the same token, many of the teachers seemed to perceive the very heart of the NMD project (merging mathematics and multiculturalism via CRP), as largely irrelevant to primary-level teaching and learning. We explore each of these ideas in the next sections.

Insistence on the Cultural Neutrality of Mathematics

D'Ambrosio (1999) characterized mathematics as one of the ethnosciences defined as "corpora of knowledge established as systems of explanations and as ways of doing, which have been accumulated through generations in *distinct natural and cultural environments* [italics in original]." He further noted, "Ethnomathematics are these corpora of knowledge derived from quantitative and qualitative practices, such as counting, weighing and measuring, comparing, sorting and classifying" (p. 35). Put differently, the majority of the skills NMD teachers routinely emphasized in their lessons can be thought of as the sine qua non of mathematics as cultural artifact (Sleeter 1997).

Concepts and skills introduced in primary classrooms serve as important prerequisite knowledge for deep mathematical understanding children need to acquire and use in higher grades and later adult life. D'Ambrosio termed this orientation to mathematics understanding as "matheracy . . . the capability of drawing conclusions from data: inferring, proposing hypotheses, . . . it is the first step toward an intellectual posture, which is almost completely absent in our school systems" (p. 35).

Our analysis of post-teaching reflections and interviews revealed the majority of teachers simply did not acknowledge cultural groundings of mathematics as a way of seeing and being in the world. For example, when describing what it means for a child to know the mathematics topic that had been taught in a given lesson, few teachers made connections to children's cultural contexts.[2] Rather, discussions were largely confined to descriptions of a child's ability to demonstrate an operation through use of manipulatives or children's abilities to explain novel ways to solve problems.

Like D'Ambrosio (1999), we found that for teachers in our study "the main importance [was] given to numeracy, or the manipulation of numbers and operations" (p. 36). We believe the reason for this is most of the teachers never moved far beyond the idea that teaching mathematics is encompassed essentially by showing and telling children how to do mathematics. In short teachers often drew upon what they knew and understood about mathematics, and sought to transfer those same understandings to their students (Tzur 2002). The challenge was for them to reconceptualize good mathematics teaching in conjunction with a culturally sensitive conception of what it means to "see" and "understand" mathematics ideas.

Traditional orientations to mathematics teaching, for example, are drawn from notions of learning premised on the idea that what children can understand is defined and bordered by what they "see"—that is, what they are shown by their teachers. From this perspective, good instruction is necessarily assumed to encompass teachers' demonstrating mathematical ideas (procedures) with the aim of children replicating and eventually internalizing them.

By contrast, NMD promoted the idea that what children "see" (defined by *their own* experiences and sense making) is what they *truly* understand. Moreover, children's own mathematical sense making represents an important filter through which teachers can most effectively help children eventually "see" (understand) new mathematics ideas (broad concepts).

To teach in this manner they had to negotiate a careful change whereby they would identify and draw on the mathematical sense makings of children. Then they needed to align those sense makings (however unconventional they may have been) with mathematics ideas the children did not yet understand but were capable of learning (Simon et al. 2000, 2004).

By the end of the study, most teachers were not teaching in this manner. Moreover, we conclude that despite the aggregate gain in their own content knowledge over the span of the project (or perhaps because of it), most of the teachers held firmly to the belief that mathematics is an objective, culturally neutral area of study. As such, few could articulate how they drew upon children's cultural backgrounds and extant sense makings to structure their teaching or learning activities to facilitate understanding of mathematics ideas.

Irrelevancy of Cultural Relevancy in Primary Grade Teaching

A number of the teachers struggled in substantive ways with the concept of cultural relevance. Based on their responses in post-teaching reflection sessions and interviews, we discovered some perceived discussions of culture in the NMD project as superfluous (if not irrelevant) to their work with kindergarten, first, and second grade children. It's possible the discomfort some teachers reported vis-à-vis the multiculturalism/CRP focus might have been owing to a perceived clash with their professional identities as primary-level teachers.

Throughout the project many alluded to children's innocence and lack of awareness of the harsh sociopolitical and economic realities of the world. Implicit in these references was the idea that adults (i.e., primary teachers) must protect young children from these realities. As the project progressed some teachers may have perceived the protector role required them to remain oblivious to the impact harsh realities present for opportunities to learn and for children's future life chances. Perhaps this also gave rise to the related idea that, except for its high academic achievement component, multiculturalism/CRP is largely irrelevant in primary grade teaching.

These erroneous perceptions are likely traceable, in part, to the inadvertent conflation of culture and race in studies of multiculturalism. For example, in an analysis of findings from a two-year professional development study, Sleeter (1997) reported teachers "did not find multicultural education useful mainly because of its stress on color, believing that acknowledging color or other ascribed characteristics would either reinforce limiting stereotypes or excuse people from trying" (p. 688). Second grade teacher Hadley shared comments reflecting aspects of this sentiment.

> I think that . . . there's no easy quick answer but it has to be where kids are motivated to learn no matter if they're Black, White, Hispanic. . . . It's not just giving them things but for them to feel pride. And I can understand how they can feel pride in a culture but I think sometimes if you focus too much on culture then . . . they know that their culture is maybe different or my culture is different. But if you focus too much on the culture then I have also seen that backfire like, well what are you going to do for me?

At the first retreat of each project year, teachers' responses to the question, *What is culture?* were telling. Many associated culture with race (color). They followed up by saying they did not see color, and the children in their classes were "colorblind" as well. Each year we challenged these ideas; however, in DCY-3, we assigned the teachers to conduct a simple activity with their own students the data from which surprised many.

Students were asked to make a sign to put outside the classroom door to let visitors know how many people were in the room. The signs created by the students showed how they saw themselves as well as their classmates. Contrary to the teachers' expectations, the children were definitely not colorblind. Rather, they were quite aware of the different skin and hair colors they saw daily among their classmates in school as these details showed up in the signs they designed (McCulloch, Marshall, and DeCuir-Gunby 2009).

Ladson-Billings (2006) wrote "Of the three terms . . . I use to describe the components of culturally relevant pedagogy, I find the notion of cultural competency the most difficult to convey to teachers who wish to develop their own practice in this way" (p. 35). She continued,

> My sense of cultural competence refers to helping students to recognize and honor their own cultural beliefs and practices while acquiring access to the wider culture, where they are likely to have a chance of improving their socioeconomic status and making informed decisions about the lives they wish to lead. The point . . . is to maintain teachers' focus on what improves the lives of the students, families, and communities they serve—not make the teachers feel better about themselves. (pp. 35–36)

We too found the cultural competence component of CRP to be the more confounding of the three, but for another reason. Different from Ladson-Billings, however, we did not place direct emphasis on teachers helping their students recognize their own cultural beliefs or negotiate mainstream culture ways as such. Instead, we emphasized teachers recognizing themselves as cultural beings. This we did through engaging them in activities where they reflected on such topics as the sources of their own value orientations, ethnoracial identity status, beliefs about what constitutes standard language, and critiques of such common terms as "culturally different" and "at risk."

An aim was to raise teachers' consciousness about how their own orientations to teaching (e.g., language usage, references to outside of school experiences, and so on), sans monitoring, could inadvertently advantage some children while at the same time disadvantage others. This focus seemed especially relevant in light of their initial claims of colorblindness.

At the same time, however, this more direct teacher-as-cultural-being focus may have been a significant departure from at least some of the teachers' own preconceptions of what constitutes appropriate professional development to integrate children's cultures into their teaching. Even teachers who eventually rejected previous claims of colorblindness (e.g., Tasha, Kendra) struggled to articulate how they drew upon their students' cultures and sense making in the context of mathematics teaching.

Only a few teachers (e.g., Johnetta, Aisha) had been initially drawn to or especially interested in the multiculturalism/CRP focus of the study, and thereby were prepared for the personal introspection required. Others did not expect and initially were not prepared for (e.g., Giselle, Arlene, Kendra)—or were offended by (e.g., Hadley)—the depth of exploration of their own cultural backgrounds and the culture/mathematics connection.

We concluded that the multiculturalism/CRP focus was likely in direct contrast to most of the teachers' own prior knowledge of and expectations for professional development focused on the place of culture in the teaching-learning process. For example, Hadley's notion of incorporating children's culture in mathematics teaching, with its human relations underpinning (Sleeter and Grant 2009), was in direct opposition to the project's focus.

> So yes we should celebrate diversity and celebrate culture and bring that into the classroom and each child no matter what color they are should feel special for who they are, where they come from and what they can add to the classroom. And as a teacher I want to make sure that I'm respecting each child. But I should not treat a Hispanic child differently than I treat a White child. I should not treat an African American child different just because of their background. But all of us are bringing things to the table. No matter what those things are, we should celebrate.

TOWARD A LIBERATING PROFESSIONAL IDENTITY: INTERSECTING CHALLENGES IN CONTEMPORARY TEACHING

The concept of intersectionality highlights how the impact of any given sociopolitical phenomenon can manifest across multiple axes (e.g., race, gender, sexuality, class, language). Studies of oppression, for example, have used intersectionality to explicate the dynamics of multiple axes converging to magnify the disadvantage and devaluation experienced by members of marked social groups (Crenshaw 1989; Lorde 1984; Stirratt et al. 2007).

In this final section, we trade on the concept of intersectionality to theorize the structure of NMD moments that supported teachers' movement toward a liberating pedagogical identity. Using experiences of two of our primary case teachers we describe how moments, which represented what we term *affirmative intersectionality*, were key to the project's overall success.

A sobering reality throughout NMD was recognizing that the teachers perceived two axes along which they would likely encounter frustration and disaffirmation were the project's actual foci. Consequently, these two highly charged areas—mathematics and multiculturalism—represented the two axes around which we structured affirmative intersectionality moments.

Throughout the project we created intentional and continuous interplay between the two. At the same time, however, we recognized these foci represented *axes of vulnerability*. Whether it was mathematics, multiculturalism, or both, each axis of vulnerability was an irritant to an individual teacher. Independently it awakened her insecurities, yet in some cases it seriously threatened her self-confidence.

Mixed Emotions about Mathematics

When teachers' insecurities, self-confidence, or even professional concerns (Marshall 1996) are activated in conjunction with a particular subject or area of study this can provoke a variety of emotions. Behaviors that accompany these emotions may range along a continuum from aggression to paralysis (Chang 2009; Schutz and Lee 2014; Zembylas 2003). Possibly more than any other subject in the contemporary curriculum, mathematics triggers emotions (often anxieties) and thereby it commonly represents an axis of vulnerability for elementary teachers in general and for many primary grade teachers in particular.

This response to mathematics may be attributed in part to the traditional orientation to elementary school teaching in which emphasis was placed on helping young students acquire basic literacy skills. Many NMD teachers saw themselves as literacy specialists. In direct contrast to their literacy expertise, however, the teachers' autobiographical narratives offered important insights into why they perceived the project's mathematics focus as having the potential to expose their weaknesses and shatter their confidence.

To be clear, a number of the teachers described positive early mathematics experiences. For instance, some reported "smooth sailing" through mathematics in both middle and high school. A number even described having successfully completed advanced mathematics courses (e.g., Calculus) during their undergraduate studies. Still, such experiences were the exception. Far more common were descriptions of underachievement, dread, and in some cases fear of mathematics as a subject of study (McCulloch et al. 2013).

Ironically, nearly every teacher (including those who were most fearful) recalled positive, affirming experiences with mathematics in elementary school. Nevertheless, some detailed how during adolescence they began to perceive they could not learn mathematics. Such self-perceptions were attributed to experiences with unkind teachers or ineffective instructional methods.

Many recalled struggling through high school mathematics courses. Some even acknowledged their successful efforts to shun and circumvent mathematics as part of their pre-professional teacher preparation. In short, it became apparent why joining NMD aggravated the vulnerability many of the teachers associated with mathematics.

Of the four primary case teachers, at the start of the project only Giselle expressed confidence in her mathematics content knowledge and teaching abilities. The other three teachers (Johnetta, Isabel, and Arlene) expressed varying degrees of discomfort, anxiety, and vulnerability at the realization that the slightness of their own content knowledge and teaching skill vis-à-vis mathematics could be exposed, and their confidence shaken, with most any given NMD activity.

Detrimental Dispositions Toward Multiculturalism

Whereas some of the teachers experienced discomfort over mathematics, even more were troubled in some manner by the study's focus on multiculturalism. "Multicultural education . . . is grounded in a vision of democracy, social justice, pluralism, and equality—ideals that have yet to be realized in U.S. society and its schools" (Sleeter 1997, p. 680). This raison d'être is indisputably sociopolitical, yet many teachers prefer to think of schools as largely untouched by this thorny realm. As such, multiculturalism has long been highly contested and resisted as an area of study in both pre-service and in-service teacher education (Gorski 2006; Marshall 2015; Vavrus 2010).

Similar to mathematics, the multiculturalism/CRP focus of the project represented a rather intense axis of vulnerability for most of the teachers. A component of the discomfort was the fact that multiculturalism/CRP centered on rejection of the premise that some children inevitably must fail in schools. Specifically, it countered the idea that only the few children who are mathematically inclined can be successful, while the rest are doomed to underachievement or failure. Correspondingly, the multiculturalism/CRP focus required teachers to denounce the fallacy that they are neutral participants (i.e., mere cogs in the system) in the operations and outcomes of contemporary schooling.

Drawing on our recollections of comments during retreats as well as the various qualitative data sources, the vulnerability some teachers associated with the multiculturalism/CRP focus likely was tied to its most fundamental component—cultural competency. It required teachers to acknowledge themselves as cultural beings before (or in conjunction with) directing their gaze at the cultural uniqueness or perceived peculiarities of the children they teach.

For some teachers, exploring the cultural self (e.g., their own worldviews, value orientations, ethnoracial identity, and so on) provoked profound unease.

Likewise, the sociopolitical consciousness component of CRP stirred troublesome emotions in that it required teachers to address issues many perceived to be beyond the purview of schools. Doing so provoked cross-cultural competence concerns (Marshall 1996) including insecurities about the substance of their knowledge of students' backgrounds.

Some teachers feared being blamed for historical injustices for which they bear no personal responsibility. One White teacher's discomfort bordered on paranoia and inadvertently exposed her entrenched negative attitudes about African Americans.

An issue that may have greatly aggravated teachers' vulnerabilities vis-à-vis multiculturalism/CRP was the invitation to contemplate their own professionalism through different eyes. Shulman (1998) put it well when he noted, "The professional must learn how to cope with those unpredictable matters and how to reflect on her own actions. Professionals incorporate the consequences of those actions into their own growing knowledge base, which ultimately includes unique combinations of theoretical and oral principles, practical maxims, and a growing collection of narratives of experience" (pp. 519–520).

Of the four primary case teachers, Johnetta was by far the most absorbed by, committed to, and critically engaged with the project's focus on multiculturalism/CRP. For the other three teachers (Giselle, Isabel, and Arlene), this focus represented a major axis of vulnerability that challenged them in different yet similarly profound ways.

Claiming Power in Moments of Affirmative Intersectionality

Each teacher risked regular experiences of unpleasant (or at least disaffirming) emotions about NMD due to the required active engagement with her axis of vulnerability. These emotions included frustration, embarrassment, or even resentment. The frequency of this threat, as well as the intensity of her vulnerability, made each teacher susceptible to making what for us would have been a very problematic decision. Each could have resigned from active engagement in retreats, or terminated her participation in the study outright. Yet none selected those options. Instead, each soldiered forward until the end and this was likely due to monitoring and structuring of their activities and experiences.

We believe the vast majority of the teachers judged NMD to be a very positive professional development experience due to frequent opportunities to experience affirmative intersectionality. Despite their experiences with considerable vulnerability, affirming moments were forged for each teacher through opportunities for substantive engagement with the "other" focal area of the study. That is, in contrast to her axis of vulnerability, each teacher also experienced an *axis of possibility*.

The axis of possibility was that project focus (mathematics or multiculturalism) through which a given teacher, on her own, perceived she would experience success, uplift, and genuine renewal. Significantly, it was also the feature of the project by which the individual teacher could *give back* through contributing to the growth and understanding of her peers.

We recognized early on that each teacher's experiences with her personal axis of possibility would need to be bolstered intermittently and over the span of her participation in NMD in order to counter the vulnerability provoked by the opposing feature. In short, it was the careful balance of activities between the two opposing axes that represented the affirmative intersectionality (see Figure 6.4).

Determining when and how to structure the intersectionality was based on a careful negotiation among several elements. These included, but were not limited to, evaluations of teachers' responses to successes and disappointments in various project contexts (e.g., retreat activities, mini-conferences, feedback from RAs); observations of their interpersonal interactions with peers (e.g., cross-cohort intersessions); and reflections on the nature of their relationships with the project research team (various contexts).

These loosely structured yet ongoing informal data gathering efforts allowed us to balance features of the project that, for whatever reasons, were perceived by individual teachers as threatening versus those that were received by those same teachers as affirming.

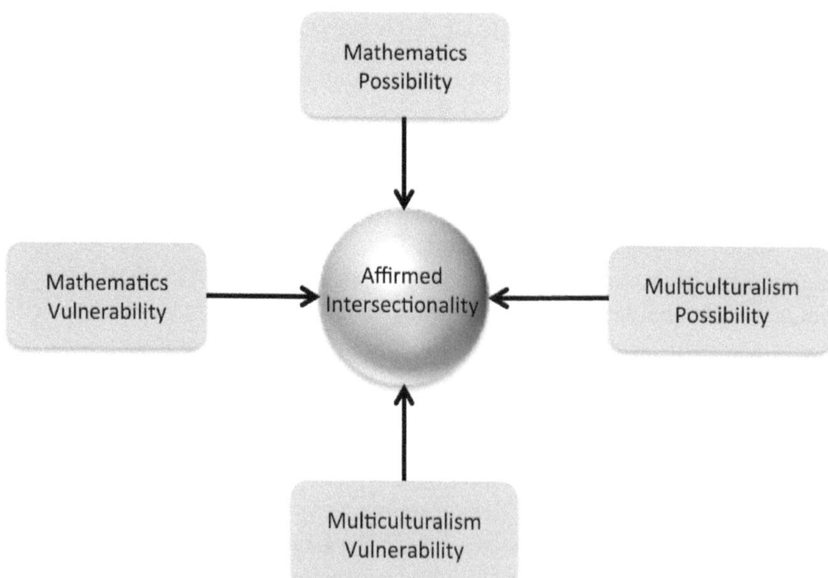

Figure 6.4 Components of Affirmed Intersectionality

The first project retreat, for example, was key in providing information about the individual teachers. It was designed for them to participate in two different activities: BaFá BaFá (Shirts 1977) and the Candy Factory. Briefly, the former is a commercial simulation designed for participants to explore the complexities of the concept "culture" along with how people react to those they perceive as "other." The latter is an interactive activity designed for teachers to deduce the essential attributes of the $base_{10}$ place value numeration system. Also, it is intended for teachers to recognize that numeration systems have cultural antecedents.

Through engagement in this activity teachers gain appreciation for the difficulties young children encounter with concepts such addition and subtraction of multidigit numbers. Both activities offered important early indications of which NMD focus would represent the axis of vulnerability versus the axis of possibility for each teacher. The responses of primary case teachers Johnetta and Giselle during and shortly after the first project retreat are illustrative.

Johnetta was thoroughly engaged in the BaFá BaFá simulation and during its debriefing session, she offered some of the most insightful and nuanced commentary of all the teachers in her cohort. She easily engaged with the project PIs over culture concepts such as the use of appropriate contemporary terminology. For instance, she expressed incredulity at the use of the term "colored children" by a White teacher, noting its usage spoke volumes about the teacher's own awareness and sensitivity.

Conversely, Johnetta's response to the Candy Factory activity was the polar opposite of what it had been for BaFá BaFá. She was quickly frustrated by its critical thinking focus, and even expressed suspicion about the point of the activity. This reaction was a useful hint at the depth of her concern about mathematics. Johnetta emerged as the primary case teacher who reacted most strongly to the possibility of failure in mathematics. She was acutely aware of the slightness of her own content knowledge, and her confidence as a mathematics learner was low.

Giselle, on the other hand, engaged enthusiastically in the Candy Factory activity and emerged as a leader of a group that offered one of the more logical and mathematically sound solutions. It didn't take long for her to see connections between the SBM focus of the project and her own enthusiastic embrace of CGI with its focus on children's active engagement in problem solving (see chapter 4).

She responded to the BaFá BaFá activity as if it were a game with no connections to real life cross-cultural interactions. A particularly conspicuous feature of her interpersonal interactions was that Giselle was not well integrated into the social milieu of the NMD teachers from her school. Early on we noticed her interactions with African American teachers were visibly strained.

Over time, we learned Giselle was perceived by some as standoffish whereas others perceived her as racially prejudiced. By her own admission, Giselle was raised with no exposure to persons of backgrounds different from her own. Initially she shut down to activities requiring her to think reflectively about her own cultural background.

Steps and Phases Along A Long Journey

Teachers joined the NMD project for various reasons. Still, we feel certain most would not have participated had they imagined the nature of its multiculturalism/CRP focus or the degree of engagement with mathematics content that was in store for them. Therefore, it was striking and ultimately consequential that depending on which NMD focal area represented the "ticket in" for an individual teacher, it also served as a significant portal through which she could traverse the borders of the perilous space represented by the other focal area. In our judgment to take on such a challenge required a disposition of fortitude. Doing the work made all the difference.

As we reflect critically on the project, we recognize our decision to combine the two axes and create moments of affirmative intersectionality served the important function of supporting each teacher's courage to delve into areas of her professional (and personal) self she unlikely would have taken on outside the NMD context. Evidence of this was especially apparent among our primary case teachers Johnetta and Giselle. In many ways they were polar opposites at the start of the project, yet they found common ground through a mutual embrace of what (in different ways) represented a liberating pedagogical identity for each of them. That identity was *culturally relevant mathematics teacher.*

While this identity was far more pronounced (across multiple data sources) among our four primary case teachers, we saw evidence of its nascent presence among other teachers as well—including among some of our secondary case teachers. Based on analysis of the various data sources, it appears teachers in our study who most successfully enacted standards-based culturally relevant mathematics teaching (i.e., embraced the identity) progressed through a series of four phases as presented in Figure 6.5.

The *Awareness* phase was probably best marked by their use of particular vocabulary associated with each project focus. Some teachers, for

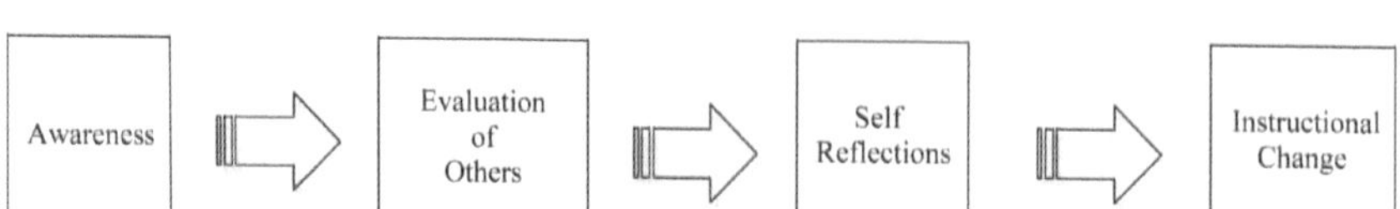

Figure 6.5 Culturally Relevant Mathematics Teacher Identity Phases

instance, adopted the term "paradigm shift" in relation to changes they saw in themselves including their emerging critical understandings of culture in the teaching-learning process (Johnetta, Kendra). Others begin using terminology we had introduced in relation to SBM, such as "procedural" versus "conceptual" teaching (Arlene).

Shortly after this phase, we observed how some teachers moved to *Evaluation of Others* followed by *Self-Reflection.* Commonly in relation to the multiculturalism/CRP focus the former phase was represented by comments about other teachers' orientations to diversity. This ranged from critiques of how other NMD teachers addressed the foci of the project (e.g., Johnetta's comments about one teacher's use of the term "colored children") to commentary on their non-NMD colleagues.

For example, after we assigned them to read the article "Peddling Poverty for Profit: Elements of Oppression in Ruby Payne's Framework" (Gorski 2008), Cohort 2 teachers from one project school (Wolfson) openly criticized the content of a workshop they had been required to attend introducing the framework. They were especially troubled by the compliant acceptance of the framework by non-NMD colleagues at their school.

Evidence of the presence of the identity in the context of *Self-Reflection* presented itself as more substantive discussions in the post-teaching reflection sessions. Initially, many teachers found it difficult to offer critical commentary to each other. By the second year, however, several remarked that they preferred when their colleagues were not trying to be "nice-nice" and "tiptoe" about errors observed in their teaching. Johnetta's comments are illustrative.

> I want to be a better educator so don't sugarcoat it because what I'm doing is impacting student learning. If I want to be a better person and my students be more successful, I need for you to be very honest in what you see because even being an experienced teacher I was still taught an old way. And because of the [Madeline] Hunter concept, the Six Point Lesson Plan, you were brought in to do this, this, this and this and not look at the culture of your students, not looking at what they say, what do you know, so all those probing questions that we started to include in the math lesson was not there initially. It was a big paradigm shift for me having been taught one way and seeing a much better way.

Instructional Change was a phase we assessed through teachers actual classroom practices. To this end, whereas the previous three phases we assessed largely through anecdotal data from project retreats, post-teaching reflections, or interviews, we elected to isolate actual teaching acts to determine how the culturally relevant mathematics teacher identity had "manifested" in practice.

In other words, we wanted to gain a rough sense of the extent to which the identity had appealed to the entire population of teachers in the study.

Therefore, for each teacher we determined the average number of tallies per lesson for the CRP subskills and for the SBM subskills as delineated in the lesson analysis rubric.

Afterward, we transferred our findings to a 3 × 3 matrix (see Figure 6.6), which provided a visual display of the extent to which each teacher had incorporated the two project foci into her actual lessons. The arrow constituting the vertical axis in the figure represents the range (from low to high) of exhibition of SBM in practice; whereas the arrow constituting the horizontal axis represents the range of exhibition (from low to high) of CRP in practice. Icons in Square 1 represent teachers who exhibited the lowest levels of SBM and CRP in the context of their lessons; whereas icons in Square 9 represent those who exhibited the highest levels of SBM and CRP in lessons.

Significantly, of the seven teachers represented by the icons presented in Square 9, over 70 percent were members of Cohort 1, whereas of the eight in Square 1 none was a member of Cohort 1. Instead, all but one (nearly 88 percent) of the teachers who exhibited the lowest levels of SBM and CRP in practice (as determined by lesson analysis) was a member of Cohort 3.

We interpreted this finding to mean that the time (i.e., number of years) a teacher participated in the NMD project was likely a significant factor in the degree to which she took on the culturally relevant mathematics teacher

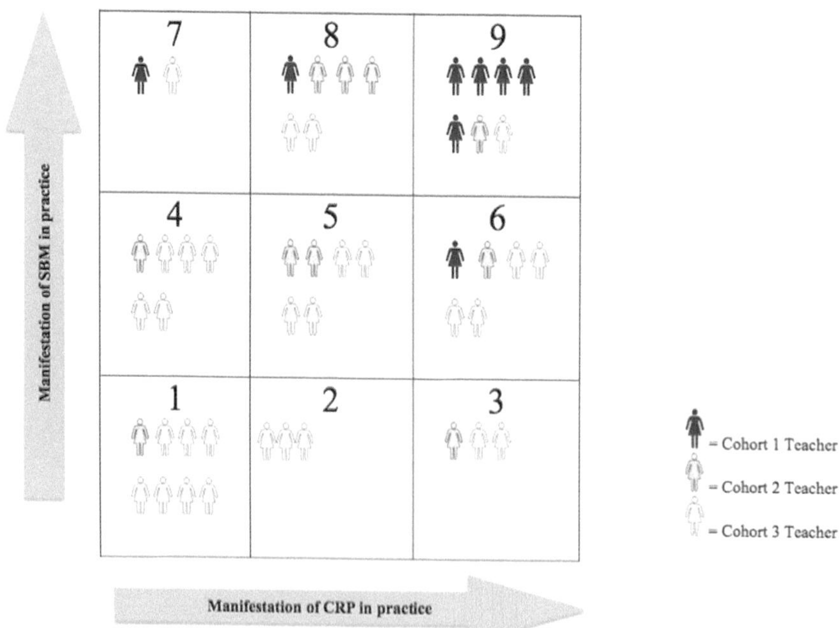

Figure 6.6 Manifestations of CRP and SBM in Practice

identity. Indeed, of the eight teachers who comprised Cohort 1, all but one exhibited moderate to high SBM and CRP in her lessons. Interestingly, nearly 45 percent of Cohort 3 teachers ranked in the moderate zone (Squares 4–6) in terms of incorporating the two emphasis in their teaching as well. To the extent we can claim matrix position is an indication of the teacher's interest in *trying on* the culturally relevant mathematics teacher identity, these moderate positions may suggest many of the teachers were at least intrigued by the idea.

CONCLUSION: CREATIVE CURIOSITY

The broad identity orientation of culturally relevant mathematics teacher was probably best encompassed in something we termed *creative curiosity*. We define this concept thusly,

> A professional identity that requires teachers to be secure in *and* uncomfortable with the extent of what they know about teaching and learning. Because of this seemingly paradoxical reality, such teachers can experience . . . what might be termed transgressive moments (hooks, 1994) in which they may "act out." It is through actual or vicarious acting out that [they] become inquisitive about possibilities in their pedagogy (curious) and they seek ways to work around or subvert business as usual teaching by becoming inventive (creative) in their own practice. (Marshall, McCulloch, and DeCuir-Gunby 2010, p. 135)

Lest we be misunderstood, we want to make clear that the claim each teacher made to the identity of *culturally relevant mathematics teacher* was not secure. On the contrary, it was emerging and at times conflicting. To be sure, we believe their claims of commitment to the identity were resolute. Yet we recognize that each teacher was also finding her way through a maze of emotions and emergent skills.

Additionally, she was exploring concepts that simultaneously frightened her and called out to her sense of obligation to do right by young students. Hence, it can be said that like an enthusiastic toddler, by the close of NMD most of the teachers remained unsteady on their feet as they took steps toward embracing a new professional identity.

Despite their unsteadiness, however, several exhibited the liberating zeal of the newly converted. Some reported efforts to share readings from the project and to open "difficult" conversations about multiculturalism with their non-NMD colleagues. Others arranged to make after-school presentations to their non-NMD colleagues that were structured around activities introduced in the project. Ultimately, the liberating element of NMD was that it allowed each K-2 teacher to try out (and to varying degrees take on) an empowering pedagogical identity. We believe it was an identity previously foreclosed to her due to an imbalance between her axis of vulnerability and her axis of possibility.

NOTES

1. Data were collected from students in the classrooms of Cohort 3 teachers and a control group of teachers. See chapter 2 for the complete discussion of this aspect of the study.

2. We acknowledge the sheer enormity of the task of learning about and integrating children's experiences into lessons. For example, Civil and Andrade (2002) found it was often difficult for teachers to extract school mathematics from the everyday lives of students even when teachers spent substantial time in the children's lives outside of school. Lest we be misunderstood, we certainly do applaud the efforts of some NMD teachers who tried to make connections for their students. At the same time, however, their successes in doing so were not commonplace and connections made were often superficial.

Appendix A

Participant Demographics

Table A.1

Category	*Cohort I 2005–2006 (DCY-1)*	*Cohort I 2006–2007 (DCY-2)*	*Cohort II 2006–2007 (DCY-2)*	*Cohort I 2007–2008 (DCY-3)*	*Cohort II 2007–2008 (DCY-3)*	*Cohort III 2007–2008 (DCY-3)*	*Control 2007–2008 (DCY-3)*
Gender	Women = 21 Men = 0	Women = 9 Men = 0	Women = 20 Men = 0	Women = 8 Men = 0	Women = 10 Men = 0	Women = 31 Men = 0	Women = 16 Men = 0
School	Farren Reese = 9 Borden = 5 Rhine River = 7	Farren Reese = 6 Rhine River = 3	Farren Reese = 9 Rhine River = 7 Wolfson = 5 Clark = 8	Farren Reese = 6 Rhine River = 2	Farren Reese = 7 Wolfson = 3	Farren Reese = 1 Rhine River = 2 Wolfson = 2 Dawson Dome = 5 Juniper = 12 Woods = 9	Highland = 16
Race	Black = 5 White = 16	Black = 3 White = 6	Black = 4 White = 15 Asian = 1	Black = 4 White = 4	Black = 4 White = 5 Asian = 1	Black = 2 White = 29	Black = 1 White = 14 Latina = 1
Grade Level	Grade K = 7 Grade 1 = 10 Grade 2 = 4	Grade K = 3 Grade 1 = 3 Grade 2 = 2 Other = 1	Grade K = 9 Grade 1 = 8 Grade 2 = 3	Grade K = 2 Grade 1 = 4 Grade 2 = 2	Grade K = 4 Grade 1 = 3 Grade 2 = 3	Grade K = 12 Grade 1 = 7 Grade 2 = 12	Grade K = 6 Grade 1 = 8 Grade 2 = 2
Avg. years at current school	4.48	3.33	5.26	4.25	6.00	3.63	1.44
Avg. years of teaching	11.81	13.78	10.1	15.06	11.40	7.98	8.81
Highest Degree	4-year = 17 Master's = 4	4-year = 8 Master's = 1	4-year = 16 Master's = 4	4-year = 7 Master's = 1	4-year = 6 Master's = 4	4-year = 23 Master's = 8	4-year = 6 Master's = 10

Note: One school (school #8) did not participate in data collection and is not included in the table.

Appendix B

NMD Mathematics Lesson Analysis Rubric

Teacher Id#: ____________ Grade: ________________ School: _______________

Lesson Date: ____________ RA/DC on Site____________ Topic(s) Taught: ________

Videotape Coder: ________ Date Coded: ____________

Directions: Use this rubric to track the nature of the teacher's verbal communication during mathematics lesson. Focusing on *teacher-initiated* verbal communication directed at an individual or group of students, codify the extent to which the teacher: 1) introduces elements of cultural relevance to the lesson and 2) attempts to illuminate and utilize students' mathematical thinking to develop/direct the lesson. The teacher's verbal communication to be codified includes instructional directives, questions about the lesson, extraneous comments related to student(s) attentiveness, behavior, and/or readiness for instruction. Using the attached sheet, collect the data while viewing the videotape. Keep a tally of the frequency with which the particular category of teacher communication occurs within each 2-minute segment of the lesson. Below is a description of each category of teacher communication to be tracked through use of this rubric.

Communication Category	Definition	Example
Learning Connecting	Teacher prompts child(ren) to recall mathematics learning as experienced in different/previous classroom-based mathematics lesson(s) taught by this teacher. Goal of communication is to help child(ren) make associations between what has been learned, and what is about to be learned.	T: In our lesson yesterday, we measured items in our classroom using string. Today, we will learn to measure these same items, but this time we'll be using different kinds of rulers.
Illuminating Thinking	Teacher provides instructional directive or poses question(s) to draw attention to and/ or highlight child(ren)'s understanding of particular mathematics topic/ concept, problem, procedure. Or, teacher poses question to allow child to more fully articulate thinking used to derive an answer. Teacher uses such terms as explain, clarify, convince me, prove, etc.	T: Shante, you said there are more girls with an odd number of pockets than there are boys. Share with us how you arrived at your answer.

Communication Category	Definition	Example
Affirming Multiple Representations	Teacher accepts children's invented written representations including pictures, charts, mathematical symbols, etc. as opposed to imposing formal mathematical symbols as a single acceptable representation	T. Okay class. We can see that Joshua drew pictures of fish, Keshia used dots, and Miguel used numbers to solve the word problem. Notice how all of these look very different, but they mean the same thing.
Extension of Task	Teacher recognizes initial task is not appropriately challenging, and in response poses a question or directive that extends the task so that the current lesson becomes challenging for a particular student, group of students, or entire class.	T. I can see that everyone understands how to add 2-digit numbers, so let me challenge you a little bit. Let's try a 3-digit number.
Language Matching	Teacher makes use of child(ren)'s own language (as opposed to standardized mathematical language) when giving an instructional directive, explanation, or following up on child's comments. Purpose is to affirm child(ren)'s own ways of articulating their understanding of the mathematics, by using child's own language.	T: Okay Carante, "we gon' put them two numbers together" and then, what we 'gon do next?
Relevance Making (Practicality Making)	Teacher makes deliberate effort to help children make association between mathematics task, concept, skill, etc. (or asks child to), and then illuminates for children (or asks child(ren) to demonstrate) *how knowing* the task, concept, skill, etc. *will be helpful* in school and/or non-school situations.	T: How many of you buy things at the "Cruizers" store just down the street? If you went to that store and wanted to buy some candy and gum, how could you use what we learned today about adding numbers?
Cultural Connecting	Teacher makes reference to child(ren)'s home/out-of-school experiences with purpose of affirming and illuminating connections to particular mathematics topic, skill, concept, being explored. Does not include attempt to highlight how knowing skills will be useful; intention is to connect and affirm child(ren)'s out-of-school experience with in-school mathematics; or connect to aspects of child's experience that is not conventionally mathematical.	T: This jar is filled with different colors of beads. Beads like these are good for putting on braids, just like the one's Jade is wearing in her hair. How many beads do you estimate are in this jar?
Communalizing	Teacher gives instructional prompt/directive intended to urge children to offer support/affirmation to their peers and to think of their personal success (i.e., completing a mathematics task, understanding a mathematics concept) as linked to that of their peers. Teachers' directive is intended to encourage students to focus on the good of the group (i.e., individual's success is related to group's success) --- not simply encouraging students to cooperate with each other.	T: Here's a challenge problem for today. Remember that our goal is to make sure that we are all successful with this challenge. If one of us is not successful, then that's a problem for all of us.

Lesson Time Laps	Learning Connecting	Illuminating Thinking	Affirming Multiple Representations	Extension of Task	Language Matching	Relevance Making	Cultural Connecting	Communalizing
00:00-02:00								
02:01-04:00								
04:01-06:00								
06:01-08:00								
08:00-10:00								
10:01-12:00								
12:01-14:00								
14:01-16:00								
16:01-18:00								
18:01-20:00								
20:01-22:00								
22:01-24:00								
24:01-26:00								
26:01-28:00								
28:01-30:00								
30:01-32:00								
32:01-34:00								
34:01-36:00								
36:01-38:00								
38:01-40:00								
40:01-42:00								
42:01-44:00								
44:01-46:00								
46:01-48:00								
48:01-50:00								
50:01-52:00								
52:01-54:00								
54:01-56:00								
56:01-58:00								

Please identify all of the following that were embedded in the lesson.

Item	Imposed on Students	Student Selected	Teacher Suggested	Not Applicable
Multiple Representations Teacher identified and/or accepted several different ways/ representations of mathematics explanation, understanding				
Use of Tools Manner of tool usage in lesson.				

Item	H	M	L
Authenticity of Mathematics. Tasks are genuine or real to life for most students in class. Encompasses activities that are conceptually challenging and require students to think mathematically (i.e., draw upon mathematics principles, universals, procedures, etc.) to find answers, solutions, clarity etc. in activity. Authenticity is reflected in ease of student completion of task, extension of activity/task, and student engagement. **High:** All students are engaged, teacher extends task for particular students, student interact/converse independently about tasks **Medium:** Most students engaged, no apparent extension of activity by teacher. Additional tasks may be posed, but not conceptually more challenging. **Low:** Most students complete task quickly and subsequently disengage as evidenced by off-task behaviors. Or, most of students' effort is devoted to non-mathematical aspects of activity such as cutting, gluing, etc.			
Overall Quality of Question Types **High Quality**: Most questions posed by teacher are intellectually engaging and reflect extension of activity rather than exclusive attention to finding answers. Questions such as: "how?", "how did you?", "please explain" "can you justify?"; "how does this compare to?" **Medium Quality**: Questioning is balanced between intellectually engaging and answer focus. **Low Quality:** Most questions posed by teacher are "answer focused queries" that require students to exhibit memory of previously learned material by recalling facts, basic concepts, and answers. E.g. "what is the answer to?" "Where is?"; "Can you recall?"			

Appendix C

Sample Items from the NMD Teacher Mathematics Questionnaire (TMQ)

<u>**Question 3**</u>

Below you see a sample of the "Hundred Board" (left) and an enlarged "piece" of it with the number **N** in the center (right). Fill in the four missing numbers (above, below, to the left, and to the right of **N**) and explain your reasons for choosing each number.

1	**2**							**9**	**10**
									20
									30
									40
91									**100**

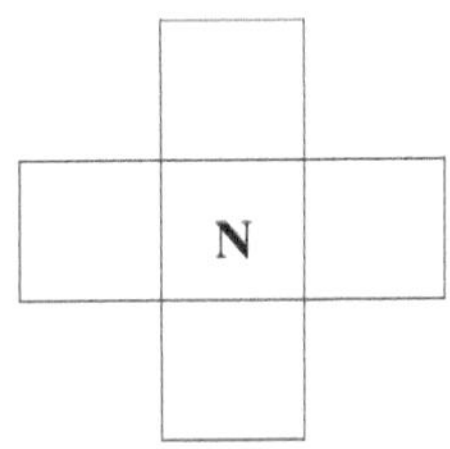

<u>Explanation</u>:

b) Where would zero be placed on the "Hundred board"? Draw a box for zero on the board and explain your reasons for choosing this location.

<u>Explanation</u>:

Question 4

Does the following strategy work for solving the subtraction problem below? Explain.

		12	16
13 ft 6 in	=>	~~13~~ ft	~~6~~ in
- 4 ft 9 in		- 4 ft	9 in
		8 ft	7 in

Answer (please circle): **1)** Yes **2)** No

Explanation:

Question 10

The following method for solving the problem 259 + 146 does **not** follow the convention of adding the "ones" first, then the "tens," and finally the "hundreds."

Will this approach **always** work? Explain.

$$259 + 100 = 359; \qquad 359 + 40 = 399; \qquad 399 + 6 = 405.$$

Answer (please circle): **1)** Yes **2)** No

Explanation:

<u>Question 19</u>

The following bar represents $1\frac{3}{5}$ of a whole. Sketch a rectangle that shows one whole. Explain

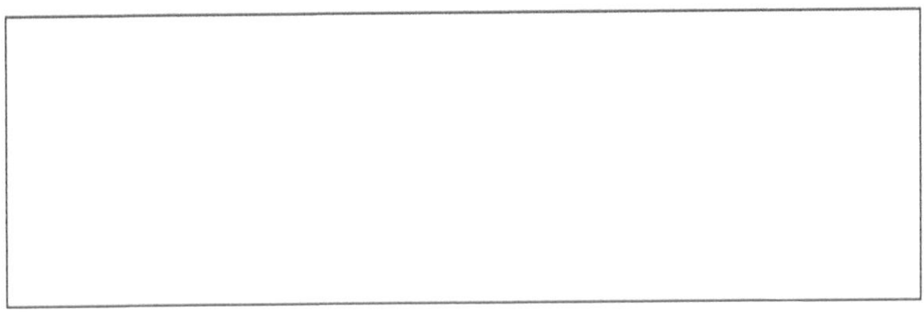

<u>Explanation</u>:

Appendix D

Teacher Dispositions Survey (TDS)

The Teacher Dispositions Survey (TDS) was designed for the NMD research project. Its purpose was to assess the nature of teachers' dispositions or orientations toward tenets of CRP vis-à-vis mathematics teaching and learning. Some items for the survey were worded in a way to "situate" CRP in the mathematics classroom context while others were worded to allude to general issues related to teaching in culturally diverse school settings. Four item categories were generated for the survey: (1) dispositions about culturally diverse students in general with some items in this category intended to assess dispositions toward working with African American students in particular; (2) dispositions about the role of parents/guardians in the teaching-learning process; (3) dispositions about teachers' need, responsibility, or inclination to affirm/value students' communities/lives outside of school; and (4) dispositions about culturally diverse school environments.

We conducted exploratory factor analyses on the instrument to test our hypothesized underlying factor structure for the scale. Several factor solutions were conducted, including three-, four-, and six-factor solutions. The three-factor solution was the best fit for the sample, explaining 33.58 percent of the variance. These three factors included *Parental Role in Schooling*— (five items); *Student Background* (five items); and *Affirm/Value Community* (ten items).

Reliability analyses indicated that the factors reflect good reliability with Cronbach's alphas of 0.931 for *Parental Role in Schooling*; 0.603 for *Student Background*; and 0.773 for *Affirm/Value Community*. Seven items were deleted from the original scale because of low factor loadings (loadings less than 0.30) or complex loadings (loadings of 0.30 or greater on 2 or more factors). In general, a higher overall score on the TDS suggested that a teacher's dispositions toward teaching in culturally diverse school settings were consistent with the thrust of the NMD research intervention. Below are interpretations of each of the three factors.

Parental Role in Schooling. Teachers whose response to the TDS reveals a positive disposition toward the *Parental Role in Schooling* factor are believed to regard the parents or guardians of their students as being genuinely interested in and supportive of their child's education. Such teachers are also believed to operate from the assumption that most parents are receptive to forming partnerships with teachers to facilitate their child's academic success, and the teachers themselves are amenable to involving parents directly in the classroom community. Less positive or negative dispositions toward this factor suggest that teachers view the parents of their students as disinterested, unhelpful, and perhaps harmful to students' academic success.

Student Background: Teachers with positive dispositions toward the *Student Background* factor regard personal characteristics of their students (e.g., cultural orientations, worldviews, learning styles and tendencies) as critical pieces of information to take into consideration in the teaching-learning process. Such teachers are believed to demonstrably incorporate aspects of students' outside-of-school lives into curricular content as well as instructional protocols with the goal of facilitating students' academic success. Teachers who hold negative dispositions toward this factor perceive the backgrounds of their students as obstacles to be overcome and/or avoided (by both teachers and students) in order for students to experience academic success.

Affirm/Value Community: A positive disposition toward the *Affirm/Value Community* factor means that teachers view the community from which their students come as an asset therefore, they encourage and attempt to demonstratively promote connections between aspects of the community and the focus of in-school curriculum and learning activities. Negative dispositions toward this factor suggest that teachers do not perceive the community (i.e., its adult members, institutions, etc.) as supportive of children's academic success. Such teachers may even perceive the community works in opposition to the school's policies and values, and that the child's outside-of-school resources (human and institutional) are ill equipped to facilitate the child's academic success.

SAMPLE TDS ITEMS

Directions: Consider each item as it relates to the teaching and learning of elementary mathematics. Circle the number that matches your belief about each item (1 = Strongly Disagree; 2 = Disagree; 3 = Neutral; 4 = Agree; 5 = Strongly Agree). If an item includes more than one aspect (indicated by a, b, c, etc.), circle a level of agreement with **each one** of them.

Table D.1 Sample Items from Teacher Dispositions Survey

#	*Survey Item*	*SD*	*D*	*N*	*A*	*SA*
1.	The "perfect mathematics classroom" would be one where all children submit their homework and are attentive to the lessons taught by their teachers.	1	2	3	4	5
2.	Most U.S. public schools have a "school culture" that is consistent with the collective White American culture.	1	2	3	4	5
3.	To foster the academic achievement of African American students in mathematics, teachers owe these students additional support due to cultural differences between the school and the students' home.	1	2	3	4	5
4.	A major problem with school mathematics is that there simply are too many children who are not ready to learn.	1	2	3	4	5
5.	Teachers should visit the homes of their students to enhance students' learning of mathematics.	1	2	3	4	5
6.	a. Parents of **African American** students want the best for their children in mathematics.	1	2	3	4	5
	b. Parents of **Asian American** students want the best for their children in mathematics.	1	2	3	4	5
	c. Parents of **Hispanic American** students want the best for their children in mathematics.	1	2	3	4	5
	d. Parents of **Native American** students want the best for their children in mathematics.	1	2	3	4	5
	e. Parents of **White American** students want the best for their children in mathematics.	1	2	3	4	5
7.	It is reasonable to expect elementary teachers to learn about the cultural backgrounds of every student in their class.	1	2	3	4	5

Appendix E

Sample Items from the NMD Student Mathematics Surveys

Kindergarten Mathematics Survey

#	Question	Records
1	a. Ask: Will you count for me? b. Please count again, this time start at 7 (--> 25) c. Please count again, this time start at 14 (-> 25) d. Please count **back** from 10	a. Correctly Reached ______ Other Count____________________ b. Started at: ___ Correct reached ____ c. Started at: ___ Correct reached ____ d. Correctly reached ______
4	a. Tell child: Please take cubes from the big pile to make a group of **8** cubes here. b. Tell child: Please tap with your finger on the desk **8** times. c. Now, count how many times I tap (up to 25). (Make sure child counts out loud)	a. 1) Accurate 2) Mistake at ______ b. 1) 1-to-1: N / Y 2) Stopped at #___ c. Correctly reached: ____
5	a. Put **7** cubes in a **line**. Ask: How many cubes are in the line? (If child does not get to **7**, re-count with them to establish the number; if needed, change to **5**) b. Re-arrange (**dense group**); Ask: How many cubes are there now? (Put cubes close, but not touching one another) c. Put out another, **more** stretched-out line with **same** number, **different color** cubes; Ask: Which line has more cubes, the (Color 1) line or the (Color 2) line or they are the same? d. If child did not count cubes, then ask: Can you count to check? Once counted, ask: So which line has more?	a. Answer: ____ 1-to-1: 1) No 2) Yes b. Answer: ____ Re-count? 1) No 2) Hesitate/Yes 3)Yes c. Answer: 1) Long 2) Short 3) Same Independent Count: 1) No 2) Yes d. Answer: 1) Long 2) Short 3) Same

1st Grade Mathematics Survey

Question 1b

Fill in the missing numbers (count by ones).

78, 79, ____, ____, ____, ____, 84,

____, ____, ____, 88, ____, ____, ____, 92,

____, ____, ____, 96, ____, ____, ____.

Question 4

The picture shows all the stars that Pat has out of her box.

She has 5 stars in the box.

How many stars will be in the box if she puts all of them in?

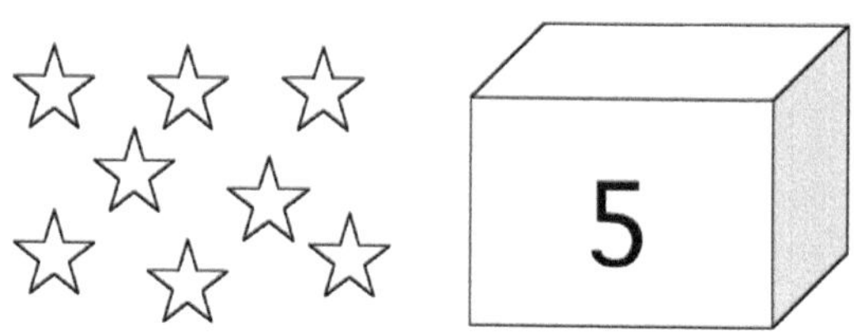

Show/explain your answer:

Question 6

The example shows a picture and a number sentence: 8+2=10. 8 shaded boxes and 2 un-shaded boxes make 10. Now you try.

Shade the other pictures to show different ways to make 10. For each picture that you create, write a number sentence that matches the picture.

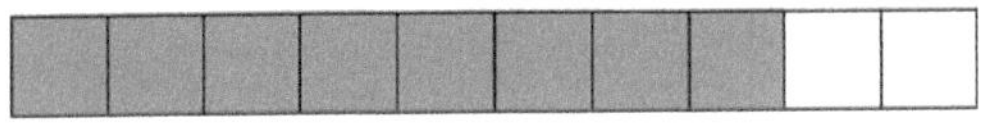

Example:

Number Sentence: 8 + 2 = 10 .

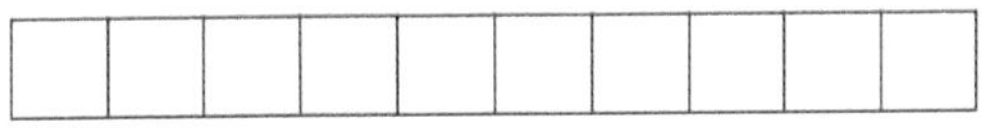

6a)

Number Sentence: ____________________.

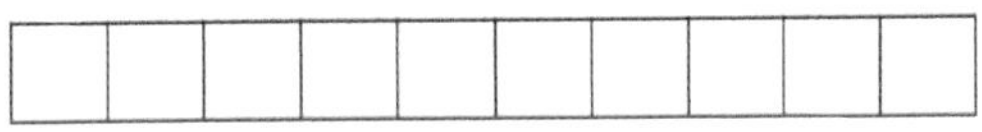

6b)

Number Sentence: ____________________.

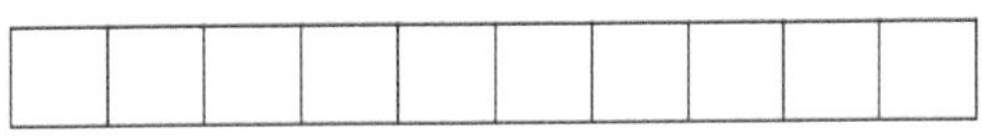

6c)

Number Sentence: ____________________.

Question 11

Julie covers the floor with small, square tiles.

How many more tiles does she need to cover the entire floor?

Show and explain ways to solve this question.

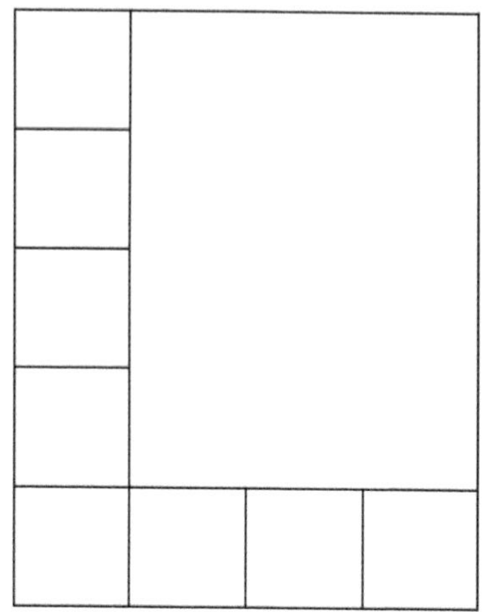

Solution 1:

Solution 2:

2nd Grade Mathematics Survey

Question 2

Using two of these cards

0 **1** **2** **3** **4**

What is the smallest 2-digit number you can make? Write the number in the boxes below.

What is the largest 2-digit number? Write the number in the boxes below.

Question 6

George has 62 pennies. Pat has 79 pennies. How many pennies does Pat need to give to George so they have the same number of pennies?

Show and explain your answer below.

Question 15

Draw a loop around $\frac{1}{2}$ of the stars below. Explain your answer.

* * *   * * *   *

* * *   * *

Appendix F

Additional Findings from the Teacher Mathematics Questionnaire

Findings related to the Teacher Mathematics Questionnaire (TMQ) (Appendix C) were reported in chapter 6. This included a comparison of the NMD teachers as a whole, Cohorts 1, 2, and 3 to the Control Group with respect to explanation scores. In addition, analysis at the item level was reported. Here we include our analysis of cohort effects on the growth of questionnaire correctness scores using ANOVA. These results appear in Table F.1 (Cohort 0 = Control Group). (Note: Homogeneity of variance assumption was checked for each model.)

Table F.1 Contrasts of Cohort Effects on Overall Growth of Correctness Scores*

Parameter	*Estimate*	*Standard Error*	*t-Value*	*Pr > |t|*
Cohorts 123–0	1.52638889	0.64889401	2.35	0.0220
Cohorts 1–23	2.86666667	0.85497557	3.35	0.0014
Cohorts 12–3	2.43333333	0.64310389	3.78	0.0004
Cohorts 12–0	2.33750000	0.74043056	3.16	0.0025
Cohorts 23–0	0.57083333	0.66515954	0.86	0.3942

*$p = 0.0008$.

All contrasts in Table F.1 are significant with the exception of the last one. Most importantly Cohorts 1, 2, and 3 together are significantly different than the Control Group. It is also of note that Cohort 1, on its own and when compared to any other pair, is significantly different. This suggests that three years in the project impacted the correctness scores more than two or fewer.

The same test was conducted with the explanation scores. The results appear in Table F.2. In addition, due to the significant increase we noted on the mean scores when the questionnaires were scored for explanations we also chose to test for individual cohort effects, which appear in Table F.3.

Table F.2 Contrasts of Cohort Effects on Overall Growth of Explanation Scores*

Parameter	*Estimate*	*Standard Error*	*t-Value*	*Pr > |t|*
Cohorts 123–0	9.3055556	1.92241395	4.84	<0.0001
Cohorts 1–23	12.6666667	2.53295136	5.00	<0.0001
Cohorts 12–3	8.6833333	1.90526014	4.56	<0.0001
Cohorts 12–0	12.2000000	2.19360022	5.56	<0.0001
Cohorts 23–0	5.0833333	1.97060223	2.58	0.0124

*$p < 0.0001$.

Table F.3 Cohort Comparisons on Overall Growth of Explanation Scores

Cohort Comparison	*Difference between Means*	*Simultaneous 95% Confidence Limits*	
1–2	11.100	3.121	19.079
1–3	14.233	7.540	20.927
1–0	17.750	10.466	25.034
2–1	-11.100	−19.079	−3.121
2–3	3.133	−3.009	9.276
2–0	6.650	−0.131	13.431
3–1	−14.233	−20.927	−7.540
3–2	−3.133	−9.276	3.009
3–0	3.517	−1.691	8.724
0–1	−17.750	−25.034	−10.466
0–2	−6.650	−13.431	0.131
0–3	−3.517	−8.724	1.691

Table F.3 shows that all contrasts are significant. Further, Cohort 1 is significantly different from any other cohort individually. Again, this suggests a benefit of a longer period of time in the intervention.

References

Apple, M. W., & Jungck, S. (1990). You don't have to be a teacher to teach this unit: Teaching, technology, and gender in the classroom. *American ucational Research Journal*, 27(2), 227–251.

Ball, D. L. (2003). *Mathematical proficiency for all students: Toward a strategic research and development program in mathematics education.* Santa Monica, CA; RAND.

Ball, D. L. (2000). Bridging practices: Interweaving content and pedagogy in teaching and learning to teach. *Journal of Teacher Education*, 51(3), 241–247.

Ball, D. L., & Cohen, D. K. (1999). Developing practice, developing practitioners: Toward a practice-based theory of professional education. In L. Darling-Hammond and G. Sykes (Ed.), *Teaching as the Learning Professions: Handbook of Policy and Practice* (pp. 3–32). Jossey-Bass: San Francisco, CA.

Ball, D. L., Lubienski, S., & Mewborn, D. (2001). Research on teaching mathematics: The unsolved problem of teachers' mathematical knowledge. In V. Richardson (Ed.), *Handbook of Research on Teaching (4th ed.).* New York: Macmillan.

Ball, D. L., Thames, M. H., & Phelps, G. (2008). Content knowledge for teaching: What makes it special. *Journal of Teacher Education*, 59(5), 389–407.

Banks, J. A. (2004). Multicultural education: Historical development, dimensions, and practice. In J. A. Banks and Cherry A. McGee Banks (Eds.), *Handbook of Research on Multicultural Education* (2nd ed., pp. 3–29). San Francisco: John Wiley &Sons/Jossey-Bass.

Baroody, A. J., & Wilkins, J. L. M. (1990). The development of informal counting, number, and arithmetic skills and concepts. In J. V. Copley (Ed.), *Mathematics in the Early Years* (pp. 48–65). Reston, VA: National Council of Teachers of Mathematics.

Borko, H. (2004). Professional development and teacher learning: Mapping the terrain. *Educational Researcher*, 33(8), 3–15.

Boykin, A. W., Albury, A., Tyler, K. M., Hurley, E. A., Bailey, C. T., & Miller, O. A. (2005). Culture-based perceptions of academic achievement among low-income elementary students. *Cultural Diversity and Ethnic Minority Psychology*, 11(4), 339–350.

Boykin, A. W., Coleman, S. T., Lilja, A. J., & Tyler, K. M. (2004). Building on children's cultural assets in simulated classroom performance environments: Research vistas in the communal learning paradigm. Center for Research on the Education of Students Placed at Risk (CRESPAR) Technical Report 68. Retrieved from: http://www.csos.jhu.edu/crespar/techReports/Report68.pdf.

Boykin, A. W., Jagers, R. J., Ellison, M., & Albury, A. (1997). Communalism: Conceptualization and measurement of an Afrocultural social orientation. *Journal of Black Studies*, 27(3), 409–418.

Brodsky, A. E. (2008). Negative case analysis. In L. M. Given (Ed.). *The SAGE Encyclopedia of Qualitative Research Methods*, (p. 553), Thousand Oaks, CA: SAGE Publications, Inc.

Brown, E. L. (2004). What precipitates change in cultural diversity awareness during a multicultural course: The message or the method? *Journal of Teacher Education*, 55(4), 325–340.

Bryk, A. S., Sebring, P. B., Allensworth, E., Luppescu, S., & Easton, J. Q. (2010). *Organizing schools for improvement: Lessons from Chicago*. Chicago: The University of Chicago Press.

Cabrera, N. L., Milem, J. F., Jaquette, O., & Marx, R. W. (2014). Missing the (student achievement) forest for all the (political) trees: Empiricism and the Mexican American studies controversy in Tucson. *American Educational Research Journal*, 51(6) 1084–1118.

Carpenter, T. P., Fennema, E., & Franke, M. L. (1996). Cognitively guided instruction: A knowledge base for reform in primary mathematics instruction. *Elementary School Journal*, 97(1), 3–20.

Carpenter, T. P., Fennema, E., Franke, M., Levi, L., & Empson, S. B. (2000). Cognitively guided instruction: A research-based teacher professional development program for mathematics. Research Report 03. Madison, WI: Wisconsin Center for Educational Research.

Chang, M. L. (2009). An appraisal perspective of teacher burnout: Examining the emotional work of teachers. *Educational Psychology Review*, 21(3), 193–218

Chinn, P. (Producer). (2004). *Culturally responsive teaching*. [Video]. National Association for Multicultural Education, Washington, DC.

Civil, M. (2002). Culture and mathematics: A community approach. *Journal of Intercultural Studies,* 23, 133–148.

Civil, M., & Andrade, R. (2002). Transitions between home and school mathematics: Rays of hope amidst the passing clouds. In G. de Abreu, A. J. Bishop, & N. C. Presmeg (Eds.), *Transitions Between Contexts of Mathematical Practices,* 149–169. Dordrecht: Kluwer.

Clements, D. H., & Sarama, J. (2007). Early childhood mathematics learning. In F. K. Lester (Ed.), *Second Handbook of Research on Mathematics Teaching and Learning* (pp. 461–556). Charlotte, NC: Information Age Publishing.

Conference Board of the Mathematical Sciences (CBMS). (2000). *The mathematical education of teachers*. Washington, DC: Mathematical Association of America.

Crenshaw, K. (1989). Demarginalizing the intersection of race and sex: A Black feminist critique of antidiscrimination doctrine, feminist theory and antiracist politics. *The University of Chicago Legal Forum*, 139–168.

Creswell, J., & Plano Clark, V. (2011). *Designing and conducting mixed methods research*. (2nd ed.). Thousand Oaks, CA: SAGE.

D'Ambrosio, U. (1999). Mathematics, history, ethnomathematics and education: A comprehensive program. *The Mathematics Educator*, 9(2), 34–36.

Darling-Hammond, L. (2004). What happens to a dream deferred? The continuing quest for equal educational opportunity. In J. A. Banks & C. A. M. Banks (Eds.), *Handbook of Research on Multicultural Education*, (2nd ed., pp. 607–630). San Francisco: Jossey-Bass.

Darling-Hammond, L., & Sykes, G. (2003). Wanted: A national teacher supply policy for education: The right way to meet the "highly qualified" teacher challenge. *Education Policy Analysis Archives*, 11, 33. Retrieved from http://epaa.asu.edu/ojs/article/view/261

DeCuir-Gunby, J. T., Marshall, P. L., & McCulloch, A. W. (2012). Using mixed methods to analyze video data: A teacher professional development example. *Journal of Mixed Methods Research*, 6(3), 199–216.

DeCuir-Gunby, J. T., Marshall, P. L., & McCulloch, A. W. (2011). Developing and using a codebook for the analysis of interview data: An example from a professional development research project. *Field Methods*, 23(2), 136–155.

Delpit, L. (2012). *Multiplication is for White people: Raising expectations for other people's children*. New York: The New Press.

Deyhle, D. (1983). Measure success and failure in the classroom: Teacher communication about tests and the understanding of young Navajo students. *Peabody Journal of Education*, 61(1), 67–85.

eMediaMillWorks (10 July 2000). *Text: George W. Bush's Speech to the NAACP*. (www.washingtonpost.com/wp-srv/onpolitics/elections/bushtext071000.htm) Downloaded 041414.

Enyedy, N., & Mukhopadhyay, S. (2007). They don't show nothing I didn't know: Emergent tensions between culturally relevant pedagogy and mathematics pedagogy. *The Journal of the Learning Sciences*, 16(2), 139–174.

Faulkner, V. M., Stiff, L. V., Marshall, P. L., Nietfeld, J., & Crossland, C. L. (2014). Race and teacher evaluations as predictors of algebra placement. *Journal for Research in Mathematics Education*, 45(3), 288–311.

Franke, M. L., Kazemi, E., & Battey, D. (2007). Understanding teaching and classroom practice in mathematics. In F. K. Lester, Jr. (Ed.). *Second Handbook of Research on Mathematics Teaching and Learning* (pp. 225–256). Charlotte, NC: Information Age Publishing.

Friedman, T. L. (2005). *The world is flat: A brief history of the twenty-first century*. New York: Farrar, Straus and Giroux.

Gay, G. (2002). Preparing for culturally responsive teaching. *Journal of Teacher Education*, 53, 106–116.

Ginsburg, H. P., Lee, J. S., & Boyd, J. S. (2008). Mathematics education for young children: What it is and how to promote it. *Social Policy Report: Giving Child and Youth Development Knowledge Away*, 22(1), 3–23.

Giroux, H. A. (2005). *Border crossings: Cultural workers and the politics of education*. 2nd edition. New York: Routledge.

Gorski, P. (2008). Peddling poverty for profit: Elements of oppression in Ruby Payne's Framework. *Equity and Excellence in Education*, 40(1), 130–148.

Gorski, P. (2006). The unintentional undermining of multicultural education: Educators at the equity crossroads. In J. Landsman & C. W. Lewis (Eds.), *White Teachers/Diverse Classrooms: A Guide to Building Inclusive Schools, Promoting High Expectations, and Eliminating Racism* (pp. 61–78). Sterling, VA: Stylus Publishing.

Gorski, P. C., & Goodman, R. D. (2011). Is there a 'hierarchy of oppression' in U.S. multicultural teacher education coursework? *Action in Teacher Education*, 33(5–6), 455–475.

Grant, C. E., & Zwier, E. (2011). Intersectionality and student outcomes: Sharpening the struggle against racism, sexism, classism, ableism, heterosexism, nationalism, and linguistics, religious, and geographical discrimination in teaching and learning. *Multicultural Perspectives*, 13(4), 181–188.

Griffin, S., Case, R., & Siegler, R. S. (1994). Rightstart: Providing the central conceptual prerequisites for the first formal learning of arithmetic to students at risk for school failure. K. McGilly (Ed.), *Classroom Lessons: Integrating Cognitive Theory and Classroom Practice* (pp.25–49). Cambridge, MA: MIT Press.

Grossman, P., Compton, C., Igra, D., Ronfeldt, M., Shahan, E., & Williamson, P. W. (2009). Teaching practice: A cross-professional perspective. *Teachers College Record*, 111(9), 2055–2100.

Gutstein, E. (2003). Teaching and learning mathematics for social justice in an urban, Latino school. *Journal for Research in Mathematics Education, 34*, 37–73.

Gutstein, E., Lipman, P., Hernadez, P., & de los Reyes, R. (1997). Culturally relevant mathematics teaching in a Mexican American context. *Journal for Research in Mathematics Education*, 28(6), 709–737.

Haberman, M. (1995). *Star teachers of children in poverty*. West Lafayette, IN.: Kappa Delta PI.

Haberman, M. (1991). The pedagogy of poverty versus good teaching. *Phi Delta Kappan*, 73(4), 290–294.

Heck, D. J., Banilower, E. R., Weiss, I. R., & Rosenberg, S. L. (2008). Studying the effects of professional development: The case of NSF's local systemic change through teacher enhancement initiative. *Journal for Research in Mathematics Education*, 39(2), 113–152.

Helms, J. E. (Ed.) (1993). *Black and white racial identity: Theory, research, and practice*. Westport, CT.: Praeger.

Hiebert, J. (1986). *Conceptual and procedural knowledge: The case of mathematics*. Hillsdale, NJ: Lawrence Erlbaum Associates.

Hiebert, J., & Carpenter, T. P. (1992). Learning and teaching with understanding. In D. A. Grouws (Ed.), *Handbook of Research on Mathematics Teaching and Learning* (pp. 65–97). New York: Macmillan.

Hiebert, J., Carpenter, T. P., Fennema, E., Fuson, K. C., Wearne, D., Murray, H., et al. (1997). *Making sense: Teaching and learning mathematics with understanding*. Portsmouth, NH: Heinemann.

Hiebert, J., & Grouws, D. A. (2007). The effects of classroom teaching on students' learning. In F. K. Lester (Ed.), *Second Handbook on Research on Mathematics Teaching and Learning* (pp. 371–404). Charlotte, NC: Information Age.

Hill, H. C., & Ball, D. L. (2004). Learning mathematics for teaching: Results from California's mathematics profesional development institutes. *Journal for Research in Mathematics Education*, 35(5), 330–351.

Hill, H. C., Rowan, B., & Ball, D. L. (2005). Effects of teachers' mathematical knowledge for teaching on student achievement. *American Educational Research Journal*, 42(2) 371–406.

Hill, H. C., Sleep, L., Lewis, J. M., & Ball, D. L. (2007). Assessing teachers' mathematical knowledge: What knowledge matters and what evidence counts? In F. K. Lester Jr. (Ed.) *Second Handbook of Research on Mathematics Teaching and Learning*, (pp. 111–156). Charlotte, NC: NCTM.

Hilliard, A., III. (2003). No mystery: Closing the achievement gap between Africans and excellence. In T. Perry, C. Steele, & A. Hilliard, III. (Eds). *Young, Gifted, and Black: Promoting High Achievement Among African American Students*. Boston: Beacon Press.

Hilliard, A., III. (1992). Behavioral styles, culture, and teaching and learning. *The Journal of Negro Education*, 61, 370–377.

Hodge, L. L., & Cobb, P. (2010, April). *Two views of culture and their implications for mathematics teaching and learning*. Paper presented at the Research Presession meeting of the National Council of Teachers of Mathematics, Denver, CO.

Hodgen, J., & Askew, M. (2007). Emotion, identity, and teacher learning: Becoming a primary mathematics teacher. *Oxford Review of Education*, 33(4), 469–487.

hooks, b. (1994). *Teaching to transgress: Education as the practice of freedom*. New York: Routledge.

Howard, G. R. (2006). *We can't teach what we don't know: White teachers, multiracial schools*. New York: Teachers College Columbia University Press.

Howard, T. C. (2003). Culturally relevant pedagogy: Ingredients for critical teacher reflection. *Theory Into Practice*, 42(3), 195–202.

Howard, T. C. (2001). Telling their side of the story: African-American students' perceptions of culturally relevant teaching. *The Urban Review*, 33(2), 131–149.

Hunt, J., & Carroll, T. (2003). *No dream denied: A pledge to America's children*. Summary report of National Commission on Teaching and America's Future, Washington.

Institute of Education Sciences (2011). *National Assessment of Educational Progress*. Retrieved from http://nces.ed.gov/nationsreportcard/.

Irvine, J. J. (1991). *Black students and school failure*. New York: Praeger.

Kazemi, E., & Franke, M. (2004). Teacher learning in mathematics: Using student work to promote collective inquiry. *Journal of Mathematics Teacher Education*, 7, 203–235.

Kazemi, E., & Stipek, D. (2001). Promoting conceptual thinking in four upper-elementary mathematics classrooms. *The Elementary School Journal*, 102(1), 59–80.

Kincheloe, J. L. (2004). *Critical pedagogy primer*. New York: Peter Lang Publishing.

Kitchener, K. S., & King, P. M. (1990). The reflective judgment model: Transforming assumptions about knowing. In J. Mezirow & Associates (Ed.). *Fostering Critical Reflection in Adults: A Guide to Transformative and Emancipatory Learning*. (pp. 159–176). San Francisco: Jossey-Bass Publishers.

Knight, S. L., & Wiseman, D. L. (2005). Professional development for teachers of diverse students: A summary of the research. *Journal of Education for Students Placed at Risk (JESPAR)*, 10(4), 387–405.

Ladson-Billings, G. (2006). Yes, but how do we do it? Practicing culturally relevant pedagogy. In J. Landsman & C. W. Lewis (Eds.), *White Teachers, Diverse Classrooms: A Guide to Building Inclusive Schools, Promoting High Expectations, and Eliminating Racism* (pp. 29–42). Sterling, VA. Stylus Publishers.

Ladson-Billings, G. (1997). It doesn't add up: African American students' mathematics achievement. *Journal for Research in Mathematics Education*, 28(6), 697–708.

Ladson-Billings, G. (1994). *The dreamkeepers. Successful teachers of African American children*. San Francisco: Jossey–Bass.

Lee, C. D. (2001). Is October Brown Chinese? A cultural modeling activity system for underachieving students. *American Educational Research Journal*, 38(1), 97–41.

Lee, C. D., Spencer, M. B., & Harpalani, V. (2003). "Every shut eye ain't sleep": Studying how people live culturally. *Educational Researcher*, 32(5), 6–13.

Leonard, J., Brooks, W., Barnes-Johnson, J., Berry, R. Q., III. (2010). The nuances and complexities of teaching mathematics for cultural relevance and social justice. *Journal of Teacher Education*, 61, 261–270.

Lincoln, Y. S., & Guba, E. G. (1985). *Naturalistic inquiry*. Thousand Oaks, CA: Sage.

Lipka, J. (2005). Math in cultural context: Two case studies of a successful culturally based math project. *Anthropology & Education Quarterly*, 36, 367–385.

Lipka, J. (1991). Toward a culturally based pedagogy: A case study of one Yup'ik Eskimo teacher. *Anthropology & Education Quarterly*, 22, 203–223.

Lorde, A. (1984). *Sister outsider: Essays and speeches*. Trumansberg, NY: Crossing Press.

Ma, L. (1999). *Knowing and teaching elementary mathematics: Teachers' understanding of fundamental mathematics in China and the United States*. Mahwah, NJ: Lawrence Erlbaum Associates.

Marshall, P. L. (2015). Using my 'you lie moment' to theorize persistent resistance to critical multicultural education. *International Journal of Multicultural Education*, 17(2), 117–134. ISSN 1934-5267. http://ijme-journal.org/ index.php/ijme/article/view/916. doi:http://dx.doi.org/10.18251/ijme.v17i2.916.

Marshall, P. L. (2013). Moodle as a data portal in a study of cultural relevance in K-2 mathematics. *The Education Forum*, 77, 277–287.

Marshall, P. L. (2009). Multicultural education in a post-race political age: Our movement at risk? *Multicultural Perspectives*, 11(4), 188–193.

Marshall, P. L. (2002). *Cultural diversity in our schools*. Belmont, CA: Wadsworth/Thomson Learning.

Marshall, P. L. (1996). Multicultural teaching concerns: New dimensions in the area of teaching concerns research? *The Journal of Educational Research*, 89(6), 371–379.

Marshall, P. L., DeCuir-Gunby, J. T., & McCulloch, A. W. (2012). In search of professional dispositions that yield cultural relevance in primary grade pedagogy: A cautionary tale of one kindergarten teacher. *Journal of Praxis in Multicultural Education,* 7(1), 1–21.

Marshall, P. L., & McCulloch, A. W. (2010). Un-bordering early mathematics pedagogy: Culture, content, and identity in critical professional development. *Bogaziçi University Journal of Education*, 27(1), 1–21.

Marshall, P. L., McCulloch, A. W., & DeCuir-Gunby, J. T. (2010). Nurturing a creative curiosity for K-2 mathematics teaching: Lessons from the dreamkeepers. In C. J. Craig & L. F. Deretchin (Eds). *Cultivating Curious and Creative Minds: The Role of Teachers and Teacher Educators. Part I. Teacher Education Yearbook XVIII.* (pp. 132–160.) Lanham: MD: Association of Teacher Educators/Rowman & Littlefield Education.

Martin, D. B. (2000). *Mathematics success and failure among African-American youth: The roles of sociohistorical context, community forces, school influence, and individual agency*. Mahwah, NJ: Lawrence Erlbaum Associates, Publishers.

Matias, C. E., & DiAngelo, R. (2013). Beyond the face of race: Emo-cognitive explorations of White neurosis and racial cray-cray. *Educational Foundations*, 27(3/4), 3–20.

McCulloch, A. W., & Marshall, P. L. (2011). K–2 teachers' attempts to connect out-of-school experiences to in-school mathematics learning. *Journal of Urban Mathematics Education*, 4(2), 44–66.

McCulloch, A. W., Marshall, P. L., & DeCuir-Gunby, J. T. (2009). Revealing cultural capital in children's number representations. *Teaching Children Mathematics*, 16(3), 184–189.

McCulloch, A. W., Marshall, P. L., DeCuir-Gunby, J. T., & Caldwell, T. S. (2013). Mathematics autobiographies: A window into teachers' identities as mathematics learners. *School Science and Mathematics*, 113(8), 380–389.

Moemeka, A. A. (1998). Communalism as a fundamental dimension of culture. *Journal of Communication*, 48(4), 118–141.

Moll, L., Amanti, C., Neff, D., & Gonzalez, N. (1992). Funds of knowledge for teaching: Using a qualitative approach to connect homes and classrooms. *Theory Into Practice*, 31(2), 132–141.

Morrison, K. A., Robbins, H. H., & Rose, D. G. (2008). Operationalizing culturally relevant pedagogy: A synthesis of classroom-based research. *Equity & Excellence in Education*, 41(4), 433–452.

Moses, R., & Cobb, C. E., Jr. (2001). *Radical equations: Math literacy and civil rights*. Boston: Beacon Press.

Murtadha-Watts, K., & D'Ambrosio, B. S. (1997). A convergence of transformative multicultural and mathematics instruction? Dilemmas of group deliberations for curriculum change. *Journal for Research in Mathematics Education*, 28(6), 767–782.

Nasir, N. S., Hand, V., & Taylor, E. V. (2008). Culture and mathematics in school: Boundaries between 'cultural' and 'domain' knowledge in the mathematics classroom and beyond. *Review of Research in Education*, 32, 187–240.

National Commission on Teaching and America's Future (2003). *No dream denied: A Pledge to America's children.* Retrieved from http://nctaf.org/research/research-papers/.

National Council of Teachers of Mathematics. (2014). *Principles to actions: Ensuring mathematical success for all.* Reston, VA: NCTM.

National Council of Teachers of Mathematics. (2000). *Principles and standards for school mathematics.* NCTM: Reston, VA.

Noguera, P. (2001). Racial politics and the elusive quest for excellence and equity in education. *Education and Urban Society*, 34(1), 18–41.

O'Day, J. A., Goertz, M. E., & Floden, R. E. (1995). *Building capacity for education reform* (CPRE Research Briefs Series, RB-18-12/95). New Brunswick, NJ: Consortium for Policy Research in Education.

Orfield, G., & Lee, C. (2006). *Racial transformation and the changing nature of segregation.* Cambridge, MA: The Civil Rights Project at Harvard University.

Patton, M. Q. (1999). Enhancing the quality and credibility of qualitative analysis. *Health Services Research*, 34(5), 1189–1208.

Paulos, J. A. (1988). *Innumeracy: Mathematics illiteracy and its consequences.* New York: Hill and Wang.

President's Council of Advisors on Science and Technology (PCAST). (2010). *Prepare and Inspire: K-12 Education in STEM for America's Future.* http://www.google.com/url?sa=t&rct=j&q=&esrc=s&source=web&cd=1&cts=1331149117620&ved=0CCYQFjAA&url=http%3A%2F%2Fwww.whitehouse.gov%2Fsites%2Fdefault%2Ffiles%2Fmicrosites%2Fostp%2Fpcast-stemed-report.pdf&ei=O7lXT8eRO8bZ0QH_s4DiDw&usg=AFQjCNFA_2H4xEWVmTlQC2K4N97mCthrKA&sig2=KN8ymv8_RdSMlrU0QdLoCA

P21 & Board, Partnership for 21st Century Skills & The Conference Board. (2006). *Are they really ready for work?* Tucson, AZ: Partnership for 21st Century Skills.

RAND Mathematics Study Panel. (2003). Mathematical proficiency for all students. Santa Monica, CA: RAND.

Sanders, W. L., & Horn, S. O. (1998). Research findings from the Tennessee value-added assessment system (TVASS) database: Implications for educational evaluation and research. *Journal of Personnel Evaluation in Education*, 12(3), 247–256.

Schutz, P. A., & Lee, M. (2014). Teacher emotion, emotional labor and teacher identity. *Utrecht Studies in Language and Communication*, 27, 169–362.

Schwartz, J. E., & Riedesel, C. A. (1994). *The relationship between teachers' knowledge and beliefs and the teaching of elementary mathematics.* Paper presented at the annual meetings of the American Association of Colleges for Teacher Education, Chicago, IL.

Shirts, R. G. (1977). *BaFá BaFá: A cross culture simulation.* [Simulation activity]. Del Mar, CA: Simulation Training Systems.

Shirts, R. G. (1969). *Star power.* [Simulation activity]. Del Mar, CA: Simulation Training Systems.

Shulman, L. (1998). Theory, practice, and education of professionals. *The Elementary School Journal*, 98(5), 511–526.

Shulman, L. S. (1986). Those who understand: Knowledge growth in teaching. *Educational Researcher*, 15(2), 4–14.

Simon, M. A., Tzur, R., Heinz, K., & Kinzel, M. (2004). Explicating a mechanism for conceptual learning: Elaborating the construct of reflective abstraction. *Journal for Research in Mathematics Education*, 35(5), 305–329.

Simon, M. A., Tzur, R., Heinz, K., & Smith, M. S. (2000). Characterizing a perspective underlying the practice of mathematics teachers in transition. *Journal for Research in Mathematics Education*, 31(5), 579–601.

Sleeter, C. E. (2008). Preparing white teachers. Teachers of color. In M. Cochran-Smith, S. Feiman-Nemser, D. John McIntyre, & K. E. Demers (Eds). *Handbook of Research on Teacher Education* (3rd edition) (pp. 559–582). New York: Routledge/Taylor & Francis Group & The Association of Teacher Educators.

Sleeter, C. E. (1997). Mathematics, multicultural education, and professional development. *Journal for Research in Mathematics Education*, 28(6), 680–696.

Sleeter, C. E., & Grant, C. A. (2009). *Making choices for multicultural education: Five approaches to race class, & gender.* (6th edition). Hoboken, NJ: John Wiley & Sons.

Sonnenschein, S., Baker, L., Moyer, A., & LeFevre, S. (2005, April). *Parental beliefs about children's reading and math development and relations with subsequent achievement.* Paper presented at the Biennial Meeting of the Society for Research in Child Development, Atlanta, GA.

Sowder, J. T. (2007). The mathematical education and development of teachers. In F. K. Lester (Ed.), *Second Handbook of Research on Mathematics Teaching and Learning* (pp. 157–224). Charlotte, NC: Information Age Publishing.

Sprinthall, N. A., & Thies-Sprinthall, L. (1983). The teacher as an adult learner. A cognitive-developmental view. In G. Griffin (Ed.). *Staff Development: Eighty Yearbook of the National Society for the Study of Education.* (pp. 13–35). Chicago: University of Chicago Press.

Stirratt, M. J., Meyer, I. H., Ouellette, S. C., & Gara, M. A. (2007). Measuring identity multiplicity and intersectionality: Hierarchical classes analysis (HICLAS) of sexual, racial, and gender identities. *Self and Identity*, 7(1), 89–111.

Tough, P. (2008). *Whatever it takes: Geoffrey Canada's quest to change Harlem and American.* New York: Houghton Mifflin Harcourt Publishing.

Tucker, C. M., & Herman, K. C. (2002). Using culturally sensitive theories and research to meet the academic needs of low-income African American children. *American Psychologist*, 57(10), 762–773.

Tzur, R. (2002). From theory to practice: Explaining successful and unsuccessful teaching activities (case of fractions). In A.D. Cockburn and E. Nardi (Ed.), *Proceedings of the 26th Annual Meeting of the International Group for the Psychology of Mathematics Education*, Norwich, UK, 297–304.

Vavrus, M. (2010). Critical multiculturalism and higher education: Resistance and possibilities within teacher education. In S. May & C. E. Sleeter (Eds.), *Critical Multiculturalism: Theory and Praxis* (pp. 19–31). New York, NY: Routledge.

Wilson, S. M., Flodin, R. E., & Fernini-Mundy, J. (2002). Teacher preparation research: An insider's view from the outside. *Journal of Teacher Education*, 53(3), 190–204.

Yin, R. K. (2004). *The case study anthology*. Thousand Oaks, CA: SAGE.

Yoon, K. S., Duncan, T., Lee, S. W. Y., Scarloss, B., & Shapley, K. (2007). Reviewing the evidence on how teacher professional development affects student achievement (Issues & Answers Report, REI. 2007- No. 033). Washington, DC: U.S. Department of Education, Institute of Education Sciences, National Center for Education Evaluation and Regional Assistance, Regional Laboratory Southwest. Retrieved from http://ies.ed.gov/ncee/edlabs.

Zembylas, M. (2003). Emotions and teacher identity: A poststructural perspective. *Teachers and teaching: Theory and practice*, 9(3), 213–238.

Index

About the Authors

Dr. Patricia L. Marshall is a professor in the Department of Teacher Education and Learning Sciences at North Carolina State University. She has developed and taught undergraduate and graduate courses in multicultural education, social justice education, school curriculum, and elementary social studies. Professor Marshall has been inducted into the North Carolina State University *Academy of Outstanding Teachers* and has been a multiple nominee for other teaching excellence honors including *Alumni Distinguished Undergraduate Professor* and the *Board of Governors Teaching Excellence Award*. Professor Marshall is listed in the Harvard Graduate School of Education Honor Roll of Teachers.

As an active member of the National Association for Multicultural Education (NAME) for over twenty years, Professor Marshall has served in various capacities, including NAME National Secretary and senior associate editor of *Multicultural Perspectives*, the official journal of NAME. She is a founding member and past president of the NC-NAME for which she received the *Pioneer Award* for outstanding commitment and service. In addition to her work with NAME, Professor Marshall is a laureate member of the Association of Teacher Educators (ATE) and has served as chairperson of the ATE Multicultural Education SIG wherein she oversaw creation and development of the newsletter, *Multicultural Signatures*. Her other work with ATE has included chairperson of the Distinguished Program in Teacher Education Committee.

Professor Marshall's research and scholarship focuses broadly on the impact of elements of culture on the teaching-learning process in U.S. precollegiate schools and teacher education. She has conducted research on teachers' concerns about working with children and youth from minoritized populations and teachers' enactment of equity pedagogies. Also, she has explored critical race theory in the social studies. Professor Marshall is coauthor on the six-volume social studies series *We the People* published by Houghton Mifflin. Venues where her writings have appeared include the *Journal of Teacher Education*, *Journal of Educational Research*, *Action in Teacher Education*, *Multicultural Perspectives*, *The Educational Forum*, *Social Studies Research and Practice,* and *International Journal of Multicultural Education*. Her book, *Cultural Diversity in Our Schools*, published by Wadsworth/Cengage, has been used widely in undergraduate and graduate courses in multicultural education.

Recently, Professor Marshall expanded her research focus to include study of the continuing education of multicultural teacher educators and later-life second language acquisition. The latter includes her own journey to acquire Spanish/English bilingualism/bi-literacy. In January 2015, she was selected for the Fulbright Scholar Core Award. She is serving as a visiting professor in the *Colegia de Ciencias Sociales* at the *Universidad San Francisco de Quito* in Cumbayá, Ecuador, where she is exploring teacher education students' investment in cross-cultural study of race and economic-class disparities in schools. Professor Marshall is a former kindergarten teacher.

Dr. Jessica T. DeCuir-Gunby is a native of Baton Rouge, Louisiana. She graduated *summa cum laude* from Scotlandville Magnet High School in 1994. She later attended Louisiana State University where she was a recipient of the AP Tureaud Scholarship Award, the LSU Outstanding Freshman Award, the Outstanding Undergraduate in Psychology Award, and was a Ronald E. McNair Scholar. In 1998, she earned her BS degree with a double major in Psychology and Spanish, graduating *summa cum laude* with Upper Division Honors in Psychology and both *Phi Kappa Phi* and *Phi Beta Kappa* distinctions. DeCuir-Gunby then attended the University of Georgia (UGA) where she was the recipient of the Gates Millennium Scholar Fellowship and a Southern Regional Education Board Dissertation Fellowship. At UGA, she earned both her MA and PhD degrees in educational psychology in 2000 and 2003, respectively.

Dr. DeCuir-Gunby is currently an associate professor of educational psychology and University Faculty Scholar in the Department of Teacher Education & Learning Sciences at NC State University. She serves as a mentor faculty with the multidisciplinary and interinstitutional Center for Developmental Science at the University of North Carolina, Chapel Hill. She is also an associate editor for the *American Educational Research Journal*. Dr. DeCuir-Gunby's research interests include race and racial identity development, critical race theory, mixed methods research, and emotions in education. Her work has been featured in top-tier journals such as *Educational Psychologist*, *Educational Researcher*, and *Review of Educational Research*. She recently served as Co-PI and methodologist on the NSF-funded grant Peer Mentoring Summits for Women Engineering Faculty of Color (ADVANCE Leadership Award Grant). The grant used mixed methods approaches, was multidisciplinary, and explored important issues in diversity and STEM.

When she is not working, Dr. DeCuir-Gunby enjoys cooking, trying out new restaurants (foodie), traveling, and spending time at home with her husband, son, and two cats.

Dr. Allison W. McCulloch grew up in Rochester, Michigan, where she graduated with honors from Rochester High School. She earned her BS in mathematics from Purdue University and MA in mathematics from Sam Houston State University. She began her career as a middle and high school mathematics teacher in Houston, Texas. During her twelve years in the classroom, Dr. McCulloch won awards at the school, district, and national levels for her teaching.

She completed her PhD in mathematics education at Rutgers, The State University of New Jersey. While at Rutgers, Dr. McCulloch was awarded a fellowship with MetroMath: The Center for Mathematics in America's Cities, an NSF-funded Center for Teaching and Learning. Through MetroMath, she worked on multiple projects with leading researchers in mathematics education and urban education from Rutgers, University of Pennsylvania, and City University of New York. This experience provided the foundation on which she has built her research program.

Dr. McCulloch is currently an associate professor of mathematics education in the Department of STEM Education at NC State University. Her research interests include the role of culture and affect in mathematics teaching and learning. She has been co-PI on the NSF-funded grant Preparing to Teach Mathematics with Technology: Expanding, Transforming, & Building Community. She is author of the book *Technology as Teddy Bear or Tool?: Students' Perceptions of Graphing Calculator Use*, and coauthor of *Preparing to Teach Mathematics with Technology: An Integrated Approach to Algebra.*

Dr. McCulloch lives in Cary, North Carolina, with her husband Ian, their two sons, Brendan and Gavin, and their two dogs. In her free time she enjoys spending time out on the lake with her family, reading, and running.

www.ingramcontent.com/pod-product-compliance
Ingram Content Group UK Ltd.
Pitfield, Milton Keynes, MK11 3LW, UK
UKHW012041150926
475560UK00001B/116

9 781475 808483